OXFORD CHEMISTRY PRIMERS

Physical Chemistry Editor
RICHARD G. COMPTON
Physical and Theoretical
Chemistry Laboratory
University of Oxford

Founding Editor and Organic
Chemistry Editor
STEPHEN G. DAVIES
The Dyson Perrins Laboratory
University of Oxford

Inorganic Chemistry Editor
JOHN EVANS
Department of Chemistry
University of Southampton

Chemical Engineering Editor
LYNN F. GLADDEN
Department of Chemical Engineering
University of Cambridge

D1145380

3 1111 01507 3321

Foundations of Science Mathematics

D. S. Sivia

Rutherford Appleton Laboratory and St John's College, Oxford

S. G. Rawlings

Department of Astrophysics and St Peter's College, Oxford

OXFORD

UNIVERSITY PRESS

*This book has been printed digitally and produced in a standard specification
in order to ensure its continuing availability*

OXFORD
UNIVERSITY PRESS

Great Clarendon Street, Oxford OX2 6DP
Oxford University Press is a department of the University of Oxford.
It furthers the University's objective of excellence in research, scholarship,
and education by publishing worldwide in
Oxford New York
Auckland Cape Town Dar es Salaam Hong Kong Karachi
Kuala Lumpur Madrid Melbourne Mexico City Nairobi
New Delhi Shanghai Taipei Toronto
With offices in
Argentina Austria Brazil Chile Czech Republic France Greece
Guatemala Hungary Italy Japan South Korea Poland Portugal
Singapore Switzerland Thailand Turkey Ukraine Vietnam

Oxford is a registered trade mark of Oxford University Press
in the UK and in certain other countries

Published in the United States
by Oxford University Press Inc., New York

© D. S. Sivia and S. G. Rawlings 1999

ISBN 978-0-19-850428-3

Printed and bound in Great Britain by CPI Antony Rowe,
Chippenham and Eastbourne

Series Editor's Foreword

Oxford Chemistry Primers are designed to provide clear and concise introductions to a wide range of topics that may be encountered by chemistry students as they progress from the freshman stage through to graduation. The Physical Chemistry series contains books easily recognised as relating to established fundamental core material that all chemists need to know, as well as books reflecting new directions and research trends in the subjects, thereby anticipating (and perhaps encouraging) the evolution of modern undergraudate courses.

In this Physical Chemistry Primer, Devinder Sivia and Steve Rawlings present an exceptionally clearly written and elegant introductory account of *Foundations of Science Mathematics*. The book explains in simple terms the basic ideas and applications of a subject which is essential knowledge for any practising scientist. This Primer will be of interest to all students of science (and their mentors).

Richard G. Compton
Physical and Theoretical Chemistry Laboratory,
University of Oxford

Preface

Mathematics plays a central role in the life of every scientist and engineer, from the earliest school days to late on into the college and professional years. While the subject may be met with enthusiasm or reluctance, depending on individual taste, all are required to learn it to some degree at university level. This Primer begins by summarising the basic concepts and results which should have been assimilated at high school, and goes on to extend the ideas to cover the material needed by the majority of undergraduates. While the emphasis may not be on the formal proof of each statement, the conveying of a meaningful understanding of where the results come from is at the heart of the book. It's hoped that the informality of the tutorial style adopted will endear the text to the novice, but that its concise nature will also ensure that it's a useful reference.

It is our pleasure to thank several friends and colleagues who provided insightful comments on a draft of the early chapters: namely, Drs. Richard Ibberson, Robert Leese, Jerry Mayers, Jeff Penfold and David Waymont. Professor Richard Compton's guidance, encouragement and patience was invaluable throughout. Finally, we gratefully acknowledge the deep influence that some mentors have had on our understanding and approach to mathematics; in particular, DSS would like to mention his father for the earliest exposure, Mr. Munson for his senior school days, and Dr. Skilling while a maturing undergraduate.

Oxford   D. S. S.
January, 1999   S. G. R.

Contents

1 Basic algebra and arithmetic

1.1 Elementary arithmetic

The first mathematical skill that we learn as children is arithmetic: starting with counting on our fingers, we soon progress to adding, taking away, multiplying, and dividing. All this seems quite straightforward until we encounter an expression like $2 + 6 \times 3$; does this mean that we add two to six and multiply the sum by three (to get 24), or should we multiply six by three and then add two (giving 20)? To avoid such ambiguities, an order ranking the priorities of the various basic arithmetical operations has been agreed; this convention can be remembered by the acronym BODMAS.

According to BODMAS, things occurring inside a bracket must be evaluated first; the next most important operation is 'of', as in half-of-six, and is rarely used; then it's division, multiplication, addition, and subtraction, in that order. Thus our simple example above can be made explicit as follows

Brackets
Of
Divide
Multiply
Add
Subtract

$$2 + 6 \times 3 = 2 + (6 \times 3) \neq (2 + 6) \times 3$$

While brackets are not necessary if the expression on the left is meant to stand for the one in the middle, they are essential for the one on the right.

Before moving on, we should also remind ourselves about the rule for multiplying out brackets; that is

$$(a+b)(c+d) = ac + ad + bc + bd \qquad (1.1)$$

where a, b, c, and d could be any arbitrary numbers (and a small space between quantities is equivalent to a multiplication, as in $ac = a \times c$).

1.2 Powers, roots and logarithms

If a number, say a, is multiplied by itself, then we can write it as a^2 and call it a-squared. A triple product can be written as a^3 and is called a-cubed. In general, an N-times self-product is denoted by a^N where the superscript, or *index*, N is referred to as the *power* of a.

$a^1 = a$
$a^2 = a \times a$
$a^3 = a \times a \times a$
$a^4 = a \times a \times a \times a$

Although the meaning of 'a to the power of N' is obvious when N is a positive integer (i.e. $1, 2, 3, \ldots$), what happens when it's zero or negative? This question is easily answered once we notice that the procedure for going from a power of N to $N-1$ involves a division by a. Thus if a^0 is a^1 divided by a, then a to the power of nought must be one (or *unity*); similarly, if a^{-1} is a^0 divided by a, then it must be equal to one over a; and, in general, a^{-N} is equivalent to the *reciprocal* of a^N. Thus, we have

$$a^0 = 1 \quad \text{and} \quad a^{-N} = \frac{1}{a^N} \qquad (1.2)$$

The basic definition of powers leads immediately to the formula for adding the indices M and N when the numbers a^M and a^N are multiplied

$$a^M a^N = a^{M+N} \tag{1.3}$$

$$a^{1/2} a^{1/2} = a$$

While our discussion has so far focused only on the case of integer powers, suppose that we were to legislate that eqn (1.3) held for all values of M and N. Then, we would be led to the interpretation of fractional powers as *roots*. To see this, consider the case when $M = N = 1/2$; it follows from eqn (1.3) that a to the power of a half must be equal to the square root of a. Extending the argument slightly, if a to the power of a third is multiplied by itself three times then we obtain a; therefore, $a^{1/3}$ must be equal to the cube root of a. In general, the pth root of a is given by

$$a^{1/3} a^{1/3} a^{1/3} = a$$

$$a^{1/p} = \sqrt[p]{a} \tag{1.4}$$

where p is an integer. One final result on powers that we should mention is

$$(a^M)^N = a^{MN} \tag{1.5}$$

which can at least be verified readily for integer values of M and N. More complicated powers can be decomposed into a series of simpler manipulations by using the rules of eqns (1.2) – (1.5). For example

$$9^{-5/2} = \frac{1}{9^{5/2}} = \frac{1}{9^{2+1/2}} = \frac{1}{9^2 9^{1/2}} = \frac{1}{81\sqrt{9}} = \frac{1}{243}$$

An alternative method of describing a number as a 'power of something' is to use *logarithms*. That is to say, if y is written as a to the power of x then x is the logarithm of y to the *base a*

$$y = a^x \quad \Longleftrightarrow \quad x = \log_a(y) \tag{1.6}$$

$$\log_{10}(10^0) = 0$$
$$\log_{10}(10^1) = 1$$
$$\log_{10}(10^2) = 2$$

$$\ln(e^0) = 0$$
$$\ln(e^1) = 1$$
$$\ln(e^2) = 2$$

where the double-headed arrow indicates an equivalence, so that the expression on the left implies the one on the right and vice versa. Since we talk in powers of ten in everyday conversations (e.g. hundreds, thousands, millions), the use of $a = 10$ is most common; this gives rise to the name *common* logarithm for $\log_{10}$, often abbreviated to just log (but this can be ambiguous). Other bases that are encountered frequently are 2 and 'e' (2.718 to 3 decimal places); we will meet the latter in more detail in later chapters, but state here that $\log_e$, or ln, is called the *natural* logarithm.

By combining the definition of the logarithm in eqn (1.6) with the rule of eqn (1.3), it can be shown that the 'log of a product is equal to the sum of the logs' (to any base); and, in conjunction with eqn (1.2), that the 'log of a quotient is equal to the difference of the logs'

$$\log(AB) = \log(A) + \log(B) \quad \text{and} \quad \log(A/B) = \log(A) - \log(B) \tag{1.7}$$

Similarly, eqn (1.6) allows us to rewrite eqn (1.5) in terms of the log of a power and to derive a formula for changing the base of a log (from a to b)

$$\log(A^\beta) = \beta \log(A) \quad \text{and} \quad \log_b(A) = \log_a(A) \times \log_b(a) \tag{1.8}$$

Thus $\ln(x) = 2.3026 \log_{10}(x)$, where the numerical prefactor is $\ln(10)$ to four decimal places, and so on.

1.3 Quadratic equations

The simplest type of equation involving an 'unknown' variable, say x, is one that takes the form $ax + b = 0$, where a and b are (known) constants. Such *linear* equations can easily be rearranged according to the rules of elementary algebra, 'whatever you do to one side of the equation, you must do exactly the same to the other', to yield the solution $x = -b/a$.

A slightly more complicated situation, which is met frequently, is that of a *quadratic* equation; this takes the general form

$$ax^2 + bx + c = 0 \qquad (1.9)$$

The crucial difference between this and the linear case is the occurrence of the x^2 term, which makes it far less straightforward to work out which values of x satisfy the equation. If we could rewrite eqn (1.9), albeit divided by a, as

$$(x - x_1)(x - x_2) = 0$$

where x_1 and x_2 are constants, then the solutions are obvious: either $x = x_1$ or $x = x_2$, because the product of two numbers can only be zero if either one or the other (or both) is nought. While such a *factorization* may not be easy to spot, it's not too difficult to rearrange eqn (1.9) into the form

$$x_1 x_2 = c/a$$

$$x_1 + x_2 = -b/a$$

$$(x + \alpha)^2 - \beta = 0$$

a procedure called 'completing the square', where α and β can be expressed in terms of the constants a, b, and c (but don't involve x). This leads to the following general formula for the two solutions of a quadratic equation

$$\alpha = \frac{b}{2a}, \quad \beta = \frac{b^2}{4a^2} - \frac{c}{a}$$

$$x = \frac{-b \pm \sqrt{b^2 - 4ac}}{2a} \qquad (1.10)$$

and is equivalent to $x = -\alpha + \sqrt{\beta}$ and $x = -\alpha - \sqrt{\beta}$. Since the square of any number, positive or negative, is always greater than or equal to zero, we require that $b^2 \geq 4ac$ for eqn (1.10) to yield 'real' values of x.

1.4 Simultaneous equations

There are many situations in which an equation will contain more than one variable; taking the case of just two, say x and y, a simple example would be $x + y = 3$. On its own, this relationship does not determine the values of x and y uniquely. Indeed, there are an infinite number of solutions which satisfy the equation: $x = 0$ and $y = 3$, or $x = 1$ and $y = 2$, or $x = p$ and $y = 3 - p$, to name but a few. To pin down x and y to a single possibility, we need one more equation to constrain them; e.g. $x - y = 1$. The two conditions can only be satisfied at the same time if $x = 2$ and $y = 1$, and are an example of solving *simultaneous* equations.

The easiest sort of simultaneous equations are linear ones; that is, where the variables only appear as separate entities to their first power (perhaps multiplied by a constant) all added together. The simplest of these is a two-by-two system, like the one above, whose general from can be written as

$$ax + by = \alpha$$
$$cx + dy = \beta$$

This can be solved by substituting for x, or y, from one of the equations into the other. For example, the first equation gives $y = (\alpha - ax)/b$; putting this in for y into the second equation yields a linear relationship for x; thus we readily obtain x, and hence y. An extension of this procedure allows us to determine the values of three variables (x, y, and z) uniquely when given three linear simultaneous equations, and so on. The only proviso is that all the equations must be genuinely distinct; that is, we cannot repeat the same one twice, or generate a third by combining any two (or more).

$$x = \frac{\alpha d - \beta b}{ad - bc}$$

$$y = \frac{\beta a - \alpha c}{ad - bc}$$

We should also note that a unique solution is only guaranteed if all the simultaneous equations are linear. If the second one in our example at the start of this section had been $x^2 - y = 3$, then the substitution of y from $x + y = 3$ would give the quadratic equation $x^2 + x - 6 = 0$. Factorization of this as $(x + 3)(x - 2) = 0$ tells us that either $x = 2$ or $x = -3$, and hence that $y = 1$ or $y = 6$ respectively.

1.5   The binomial expansion

It is easily shown, by putting $c = a$ and $d = b$ in eqn (1.1), that the square of the sum of two numbers can be written as

$$(a + b)^2 = a^2 + 2ab + b^2$$

Multiplying this by the sum again yields the following for the cube

$$(a + b)^3 = a^3 + 3a^2 b + 3ab^2 + b^3$$

When this procedure is repeated many times, a systematic pattern emerges for the Nth power of $a + b$; it is called the *binomial* expansion

$$(a + b)^N = \sum_{r=0}^{N} {}^N C_r \, a^{N-r} \, b^r \tag{1.11}$$

$$\sum_{r=0}^{N} A_r = A_0 + A_1 + A_2 + \cdots + A_N$$

The capital Greek symbol Σ stands for a 'sum from $r = 0$ to $r = N$', through the integers $r = 1, 2, 3, \cdots, N - 1$. The binomial coefficients ${}^N C_r$, which are often denoted by an alternative long bracket notation, are defined by

$${}^N C_r = \begin{pmatrix} N \\ r \end{pmatrix} = \frac{N!}{r!\,(N-r)!} \tag{1.12}$$

where the *factorial* function is given by the product

$$N! = N \times (N-1) \times (N-2) \times \cdots \times 3 \times 2 \times 1 \tag{1.13}$$

```
              1
           1     1
        1     2     1
     1     3     3     1
  1     4     6     4     1
1     5    10    10     5     1
1  6  15    20    15   6   1
```

Although not obvious from eqn (1.13), we will see later that $0! = 1$.

Another way of evaluating the coefficients in a binomial expansion, rather than using eqn (1.12), is provided by *Pascal's triangle*. In this, apart from the ones down the edges, each number is generated by adding the two closest neighbours in the line above. The third row of Pascal's triangle gives the coefficients for the square, the fourth row corresponds to the cube, and so on.

1.6 Arithmetic and geometric progressions

Sometimes we need to work out the sum of a sequence of numbers that are generated according to a certain algebraic rule. The simplest example is that of an *arithmetic progression*, or AP, where each term is given by the previous one plus a constant

$$a + (a+d) + (a+2d) + (a+3d) + \cdots + (l-d) + l$$

If there are N terms, then the last one, l, is related to the first, a, through the *common difference*, d, by $l = a + (N-1)d$. The addition of the above series with a copy of itself written in reverse order shows that twice the sum which we seek is equal to N times $a+l$; hence, the formula for the sum of an AP is

$$1 + 2 + 3 + \cdots + N = \frac{N(N+1)}{2}$$

$$\sum_{j=1}^{N} a + (j-1)d = \frac{N}{2}[2a + (N-1)d] \qquad (1.14)$$

Another case which is met frequently is that of a *geometric progression*, or GP, where each term is given by the previous one times a constant

$$a + ar + ar^2 + ar^3 + \cdots + ar^{N-2} + ar^{N-1}$$

If we subtract from the above series a copy of itself that has been multiplied by the *common ratio* r, then it can be shown that $1 - r$ times the sum that we seek is equal to $a - ar^N$; hence, the formula for the sum of a GP is

$$\sum_{j=1}^{N} ar^{j-1} = \frac{a(1-r^N)}{1-r} \qquad (1.15)$$

If the *modulus* of the common ratio is less than unity, so that $-1 < r < 1$, then the sum of the GP does not 'blow up' as the number of terms becomes infinite. Indeed, since r^N becomes negligibly small as N tends to infinity, eqn (1.15) simplifies to

$$1 - \frac{1}{2} + \frac{1}{4} - \frac{1}{8} + \cdots = \frac{2}{3}$$

$$\sum_{j=1}^{\infty} ar^{j-1} = \frac{a}{1-r} \quad \text{for} \quad |r| < 1 \qquad (1.16)$$

1.7 Partial fractions

For the final topic in this chapter, we turn to the subject of *partial fractions*. These are best explained by the use of specific examples, such as

$$\frac{5x+7}{(x-1)(x+3)} = \frac{3}{(x-1)} + \frac{2}{(x+3)}$$

The act of combining two, or more, fractions into a single entity is familiar to us, through 'finding the lowest common denominator', and is usually referred to as 'simplifying the equation'. Here we are interested in the reverse procedure of decomposing a fraction into the sum of several constituent parts. Although this may sound like a backwards step, it sometimes turns out to be a very useful manipulation.

$$\frac{1}{2} - \frac{1}{3} = \frac{3}{6} - \frac{2}{6} = \frac{1}{6}$$

The situations in which partial fractions arise involve the ratios of two *polynomials* (the sums of positive integer powers of a variable, like x), where the denominator can be written as the product of simpler components. The first

$$\begin{array}{r} 2 \\ x^2 + 2x - 3 \overline{)\,2x^2 + 9x + 1} \\ 2x^2 + 4x - 6 \\ \hline 5x + 7 \end{array}$$

thing we must check is that the *degree* (or highest power of x) of the numerator is less than that of the denominator, and 'divide out' if it is not; this is analogous to expressing a top-heavy fraction as whole number plus a proper fraction (e.g. $7/3 = 2 + 1/3$). If the numerator in our earlier example had been $2x^2 + 9x + 1$, then we would divide it by the denominator $x^2 + 2x - 3$ (when multiplied out), by 'long division' if necessary, to obtain

$$\frac{2x^2 + 9x + 1}{(x-1)(x+3)} = 2 + \frac{5x + 7}{(x-1)(x+3)}$$

The second part on the right can then be split up into partial fractions, as indicated at the start of this section; let's consider the mechanistic details of how this is done.

Our example constitutes the simplest case, where the denominator is a product of linear terms. This can always be expressed as

$$\frac{5x + 7}{(x-1)(x+3)} = \frac{A}{(x-1)} + \frac{B}{(x+3)}$$

where A and B are constants. If there had been a third term on the bottom, like $2x + 3$, then we would get an additional fraction on the right, $C/(2x + 3)$. To evaluate A and B (and C etc.), we can rearrange the right-hand side to have the same common denominator as on the left, and then equate the numerators

$$5x + 7 = A(x+3) + B(x-1)$$

$$A + B = 5$$
$$3A - B = 7$$

For this to be satisfied, the coefficients of the various powers of x on both sides must be identical; this leads to a set of simultaneous equations for the desired constants. An alternative procedure is to substitute specific values of x into the equation, especially those that make the factors in the denominator equal to zero; thus putting $x = 1$ gives $4A = 12$, and setting $x = -3$ gives $4B = 8$. The great ease of the second method leads to a short cut way of implementing it known as the 'cover-up' rule: to evaluate A, cover up $x - 1$ in the original expression and substitute $x - 1 = 0$ in what remains; for B, cover up $x + 3$ and put $x + 3 = 0$ in the rest.

If there had been a quadratic factor in the denominator of our example, then its partial component would require a linear term in its numerator

$$\frac{5x + 7}{(x-1)(x^2 + 4x + 3)} = \frac{A}{(x-1)} + \frac{Bx + C}{(x^2 + 4x + 3)}$$

Apart from this generalisation, that an Nth degree factor on the bottom needs an $(N-1)^{\text{th}}$ degree polynomial on the top, the procedure for calculating the associated constants is the same as before. Rearrange the right-hand side to have the same common denominator as the left, and then equate numerators

$$5x + 7 = A(x^2 + 4x + 3) + (Bx + C)(x - 1)$$

$$A + B = 0$$
$$4A - B + C = 5$$
$$3A - C = 7$$

In this case, A is evaluated most easily by substituting $x = 1$, giving $8A = 12$, and is equivalent to using the cover-up rule. B and C can then be ascertained readily by equating the coefficients of the x^2 and x^0 terms on both sides of the equation (and confirmed by those of x^1).

The final circumstance of note concerns the occurrence of repeated factors in the denominator, such as $(x + 3)^2$. We could expand this as $x^2 + 6x + 9$ and use the procedure for dealing with a quadratic discussed above, but a more useful decomposition tends to be

$$\frac{5x+7}{(x-1)(x+3)^2} = \frac{A}{(x-1)} + \frac{B}{(x+3)} + \frac{C}{(x+3)^2}$$

The related constants can be evaluated in the usual manner, by equating the numerators after the denominators have been made equal

$$5x+7 = (x+3)[A(x+3)+B(x-1)]+C(x-1)$$

A and C are given immediately by putting $x = 1$ and $x = -3$ or, equivalently, by the cover-up rule; B then follows for the coefficients of x^2, or x^1, or x^0.

Exercises

1.1 Evaluate (i) $1+2 \times 6-3$, (ii) $3-1/(2+4)$, (iii) $2^3 - 2 \times 3$.

1.2 Simplify eqn (1.1) when (i) $b = 0$, (ii) $a = c$ and $b = d$, (iii) $a = c$ and $b = -d$.

1.3 Evaluate (i) $4^{3/2}$, (ii) $27^{-2/3}$, (iii) $3^2 \, 3^{-3/2}$, (iv) $\log_2(8)$, (v) $\log_2(8^3)$.

1.4 By using the definition of a logarithm in eqn (1.6), and substituting $A = a^M$ and $B = a^N$, show that taking $\log_a$ of both sides of eqn (1.3) leads to eqn (1.7); similarly, show that eqn (1.5) yields eqn (1.8).

1.5 By dividing eqn (1.9) by a and completing the square, derive eqn (1.10).

1.6 Solve (i) $x^2 - 5x + 6 = 0$, (ii) $3x^2 + 5x - 2 = 0$, (iii) $x^2 - 4x + 2 = 0$.

1.7 For what values of k does $x^2 + kx + 4 = 0$ have real roots?

1.8 Solve the following simultaneous equations:

(i) $\begin{array}{l} 3x+2y = 4 \\ x - 7y = 9 \end{array}$, (ii) $\begin{array}{l} x^2 + y^2 = 2 \\ x - 2y = 1 \end{array}$, (iii) $\begin{array}{l} 3x + 2y + 5z = 0 \\ x + 4y - 2z = 9 \\ 4x - 6y + 3z = 3 \end{array}$

1.9 Expand $(2+x)^5$, $(1+x)^9$, and $(x+2/x)^6$ in powers of x.

1.10 Derive the formulae for the sums of an AP and GP.

1.11 By expressing a recurring decimal number as the sum of an infinite GP, show that $0.12121212\cdots = 4/33$. What is $0.318181818\cdots$ as a fraction?

$$3.33\cdots = 3 + 0.3 + 0.03 + \cdots$$

1.12 Decompose the following into partial fractions:

(i) $\dfrac{1}{x^2 - 5x + 6}$, (ii) $\dfrac{x^2 - 5x + 1}{(x-1)^2(2x-3)}$, (iii) $\dfrac{11x+1}{(x-1)(x^2 - 3x - 2)}$

1.13 Evaluate:

(i) $\displaystyle\sum_{n=1}^{\infty} \frac{1}{n(n+1)}$, (ii) $\displaystyle\sum_{n=0}^{\infty} e^{-\beta(n+1/2)}$

2 Curves and graphs

2.1 Straight lines

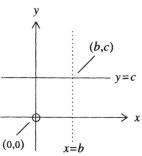

The relationship between two quantities, generically called x and y, is often best understood when displayed in the form of a *graph*. That is when linked pairs of x and y values are plotted as distinct points on a piece of paper, and joined together to form a curve. The horizontal and vertical displacements, relative to a reference location for $x = 0$ and $y = 0$ known as the *origin*, give the *coordinates* (x,y) of the points; by convention, x increases from left to right and y from bottom to top. With such a visual manifestation, it is easy to see (literally) how y changes with x. The simplest variation is none at all! This corresponds to the equation $y = c$, and yields a graph consisting of just a horizontal line whose vertical position is determined by the constant c.

A more general case of a straight line is one that is tilted. This takes the algebraic form

$$y = mx + c \tag{2.1}$$

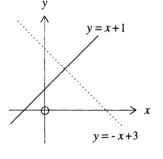

where m is a constant that controls the magnitude and sense of the slope. If m is positive, then y increases with x; if it's negative, then y decreases with x; and if $m = 0$, then we return to the situation of no variation. The values of m and c can be obtained from any two points, (x_1, y_1) and (x_2, y_2), on the line

$$m = \frac{y_2 - y_1}{x_2 - x_1} \quad \text{and} \quad c = \frac{x_2 y_1 - x_1 y_2}{x_2 - x_1}$$

and are usually referred to as the *gradient* and *intercept*, respectively. The latter is also equal to the value of y when $x = 0$, so that changing c moves the line up and down; the corresponding point on the x-axis, when $y = 0$, is called the *abscissa*.

In the light of this discussion, we can interpret the (unique) solution of the 2×2 set of linear simultaneous equations in section 1.4 as being given by the intersection of two straight lines.

2.2 Parabolas

In chapter 1, we came across quadratic equations; these involved an x^2 term, in addition to the linear ones. Their general form can be written as

$$y = ax^2 + bx + c \tag{2.2}$$

and trace out a *parabola*, like the trajectory of a cannon ball, when plotted as a graph. The sign of the x^2 coefficient, a, determines whether the curve bends upwards or downwards at the ends, and its magnitude controls how quickly it does so. To understand how a, b, and c affect the location of the parabola, it is best to rewrite eqn (2.2) by 'completing the square' as in section 1.3

$$y = a(x + \alpha)^2 + \gamma$$

It is then easy to see that the *turning point* of the curve (the apex where it bends over) occurs at $x = -\alpha$, or a half of minus b/a, and $y = \gamma$, which is c minus a quarter of b^2/a.

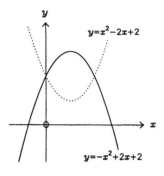

When solving a quadratic equation, we seek the values of x which make eqn (2.2) equal to zero. Graphically, these corresponds to the two points where the parabola cuts the x-axis; they become coincident when $b^2 = 4ac$, however, because the turning point then occurs at $y = 0$. There are situations when the curve never crosses the x-axis, of course, so that no (real) solutions exist; this is the case of section 1.3 when $b^2 < 4ac$.

The nature of eqn (2.2) also explains why we don't obtain a unique answer for a 2×2 set of simultaneous equations when one of them is quadratic and the other linear: we're looking for the intersections between a parabola and a straight line.

2.3 Polynomials

Both the straight line and the parabola are special cases of a family of curves known as *polynomials*. The similarity between them is more obvious from the algebraic form of the equation that defines them

$$y = a_0 + a_1 x + a_2 x^2 + a_3 x^3 + \cdots + a_N x^N \qquad (2.3)$$

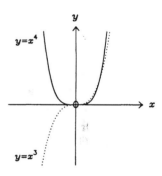

where a_0, a_1, a_2, $\cdots$, a_N, are constants; the highest power of x, N, is known as the degree, or order, of the polynomial. If all the coefficients except a_0 are zero, or put simply $N = 0$, we obtain a horizontal line; with $N = 1$, we recover the two components needed for a general straight line; and a parabola emerges if $N = 2$.

After the linear case, the graphs of the polynomials start to develop bends and wiggles in them. At the ends, where the magnitude of x is very large, all the terms in eqn (2.3) become negligible in comparison to the last one. In the extremities, therefore, the polynomial curves upwards or downwards with the same sense if the degree N is even, and in opposing ways if N is odd; while the amount of the curvature increases with N, its absolute rate and direction is governed by the size and sign of the coefficient a_N respectively. On the journey from its starting point (when $x \to -\infty$) to its final destination ($x \to \infty$), the y coordinate can go through a number of oscillations at intermediate values of x: an Nth order polynomial can have up $N - 1$ turning points, but it may have fewer by multiple factors of two. Thus, for example, a *cubic* ($N = 3$) can either have two turning points (a 'maximum' and a 'minimum') or none at all; and a *quartic* ($N = 4$) may have three or just one.

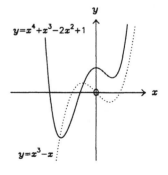

The values of x which make eqn (2.3) equal to zero are called the *roots* of the polynomial. Graphically, they correspond to the points where the curve crosses the x-axis. Given the discussion above, a little thought shows that a polynomial of degree N can have up to N 'real' roots.

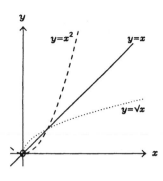

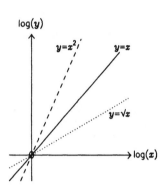

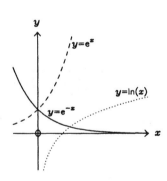

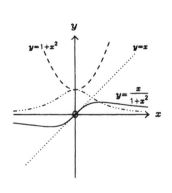

2.4   Powers, roots and logarithms

One of the most basic relationships between two quantities is that of a *proportionality*. If y becomes twice as large when x doubles, and three times as big when x trebles, and so on, then we say that 'y is directly proportional to x' and write it as $y \propto x$. This can be turned into an equation by introducing a multiplicative constant k, so that $y = kx$, and is represented graphically by a straight line that passes through the origin. Generalising from the linear case, y may be proportional to any power of x

$$y \propto x^M \qquad \Longleftrightarrow \qquad y = kx^M \qquad (2.4)$$

If $M = 2$, for example, then y increases four-fold when x doubles, and nine-fold when x trebles; by contrast, if $M = 1/2$, y is only twice as large when x is four-times bigger, and merely trebles when x increases nine-fold.

Although the algebraic form of eqn (2.4) is very simple, it's not easy to ascertain the values of M and k from a graph of y against x; this is because, other than when M is 0 or ± 1, it's difficult to judge by eye a curve that has a varying degree of bend. We can make the analysis much more straightforward, however, by taking the logarithm of eqn (2.4)

$$\log(y) = M\log(x) + \log(k) \qquad (2.5)$$

where we have used the results from eqns (1.7) and (1.8) to expand $\log(kx^M)$ on the right-hand side. Thus if we plot $\log(y)$ against $\log(x)$, to any base, for x (and k) greater than zero, we will obtain a straight line whose gradient and intercept are given by M and $\log(k)$ respectively.

Another power-related equation which crops up frequently in theoretical work is that of an *exponential decay*

$$y = A\mathrm{e}^{-\beta x} = A\exp(-\beta x) \qquad (2.6)$$

where 'exp' is an alternative notation for 'e to the power of', and A and β are constants. Strictly speaking, β must be positive for a decay; otherwise, there is exponential growth. As in the earlier situation, eqn (2.6) can be turned into a simpler graphical form by taking its logarithm (preferably to base e)

$$\ln(y) = \ln(A) - \beta x \qquad (2.7)$$

where we've used the definition of $\log_e$ from eqn (1.6) in writing the last term. In other words, if we plot $\ln(y)$ against x (for $A > 0$) then we obtain a straight line with a gradient and intercept given by $-\beta$ and $\ln(A)$ respectively.

The graphs of complicated functions of polynomials, and exponentials, can often be sketched by first considering the behaviours of their constituent parts. For example, $y = x/(1 + x^2)$ can be thought of in the following way: (i) $1 + x^2$ is a parabola that is large and positive when $x \to \pm\infty$, and has a minimum at $x = 0$ and $y = 1$; (ii) its reciprocal decays away to zero in the tails, therefore, from a (rounded) maximum value of unity at $x = 0$; (iii) multiplying this by x, or a straight line through the origin, we see that $x/(1 + x^2)$ rises to a maximum from $(0,0)$ before dying away (like $1/x$) to $y = 0$ as $x \to \infty$, and is the negative mirror image for $x < 0$.

2.5 Circles

The most perfect curve of all is probably a circle. Mathematically, it's defined as 'the locus of a point such that its distance from a fixed point is constant'. We can turn this formal jargon into an equation through a simple geometrical argument. Suppose that (x_0, y_0) is the coordinate of the fixed point, called the *centre*, and that (x, y) is some arbitrary point on the *circumference*. Then, by constructing a right-angled triangle and using *Pythagoras' theorem*, that 'the square of the hypotenuse is equal to the sum of the squares of the other two sides', we find that the distance from (x_0, y_0) to (x, y) is given by the square root of $(x - x_0)^2 + (y - y_0)^2$. Since this has to be a constant for all points on the rim, we must have

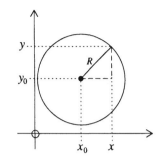

$$(x - x_0)^2 + (y - y_0)^2 = R^2 \qquad (2.8)$$

where R is the *radius* of the circle. If the centre happens to be at the origin, then eqn (2.8) reduces to $x^2 + y^2 = R^2$.

The equation above can easily be expanded and rearranged into a 'standard', if less obvious, form

$$x^2 + y^2 + 2gx + 2fy + c = 0 \qquad (2.9)$$

where the centre is at $(-g, -f)$, and the radius is given by the square root of $g^2 + f^2 - c$. For completeness, we should also add that the circumference of a circle is equal to $2\pi R$ (the definition of π) and that its area is πR^2.

2.6 Ellipses

If a circle is squashed so that it's shorter in one direction, and longer at right-angles to it, then an *ellipse* is obtained. The simplest form of the equation that describes this situation is

$$\frac{x^2}{a^2} + \frac{y^2}{b^2} = 1 \qquad (2.10)$$

It corresponds to the case where the centre is at the origin, and the *principal axes* are along those of the x and y coordinates; the widths in these directions are given by $2a$ and $2b$. We recover the circle if $a = b$, of course, so that the denominators on the left-hand side are equal to R^2. If the centre was at (x_0, y_0), then x would be replaced by $(x - x_0)$ and y by $(y - y_0)$ in eqn (2.10); we could also write a formula very similar to eqn (2.9), but the coefficients of the x^2 and y^2 terms would be different. If the *major* (or long) and *minor* (short) axes did not lie along the x and y coordinate directions, then there would be an additional xy cross-term. While it can be shown that the area of the ellipse in eqn (2.10) is $\pi a b$, there is no simple formula for its perimeter.

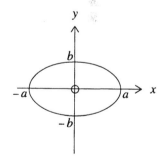

Both the ellipse and the parabola are, in fact, linked through the traditional topic of *conic sections*. That is to say, if an ordinary cone is cut in various ways, then depending on the direction of the slice, the resulting cross-section is either an ellipse, or a parabola, or a *hyperbola*. Mathematically, all three correspond to the path of a point P, with coordinates (x, y), which moves such that the ratio of its distance r from a fixed point S (called the *focus*) to its perpendicular distance l from a given reference line (known as the *directrix*) is constant; so,

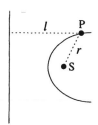

defining the latter to be the eccentricity ε, we have

$$\frac{r}{l} = \varepsilon$$

If $\varepsilon < 1$, then we obtain an ellipse; if $\varepsilon = 1$, we get a parabola; and $\varepsilon > 1$ gives a hyperbola. In terms of eqn (2.10), S is at $(a\varepsilon, 0)$ and the directrix is the line $x = a/\varepsilon$; the symmetry of the situation means that $(-a\varepsilon, 0)$ and $x = -a/\varepsilon$ would be an equally valid choice. It can be shown further that a, b and ε obey the relationship $b^2 = a^2(1 - \varepsilon^2)$, so that a circle is the special case of $\varepsilon = 0$. The orbital path of a planet is a physical example of an ellipse, with the sun at one focus, as was discovered by Kepler in 1609.

We note in passing that the equation of a hyperbola is very similar to eqn (2.10) except for a minus sign between the two terms on the left-hand side; it gives an open curve (two symmetrical ones actually), rather like a parabola, that asymptotically approaches the straight lines $y = \pm bx/a$ as x becomes very large.

Exercises

2.1 On the same graph, draw the three lines $y = x$ and $y = x \pm 1$; similarly, draw the six lines $y = \pm 2x$, $y = \pm x$, and $y = \pm x/2$.

2.2 Find the equation of the straight line that passes through the two points $(-1, 3)$ and $(3, 1)$; where does it intersect with $y = x + 1$?

2.3 Sketch the pairs of parabolas $y = x^2 \pm 1$, $y = \pm x^2 + 1$, and $y = \pm 2x^2 + 1$.

2.4 By 'completing the square', find the coordinates of the turning point of $y = x^2 + x + 1$; hence sketch the parabola.

2.5 Find the equation of the parabola that passes through the three points $(0, 3)$, $(3, 0)$, and $(5, 8)$; what are the roots of the equation?

2.6 Where does the curve $y = (x - 3)(x - 1)(x + 1)$ cross the x and y axes? Hence sketch this cubic function, and state the ranges of x values for which it is greater than zero.

2.7 Sketch the functions $y = 1 - e^{-x}$ and $y = 1 - e^{-2x}$ for positive values of x; and $y = e^{-|x|}$, $y = 1/x$, and $y = 1/(x^2 - 1)$ for all x.

2.8 Given that the exponential dominates for large x, sketch the functions $y = xe^{-x}$, $y = xe^{-2x}$, and $y = x^2 e^{-x}$ for $x \geq 0$.

2.9 What is the centre and radius of the circle $x^2 + y^2 - 2x + 4y - 4 = 0$?

2.10 Sketch the ellipse $3x^2 + 4y^2 = 3$; evaluate its eccentricity, and indicate the positions of the focus and directrix.

2.11 Sketch the hyperbola $x^2 - y^2 = 1$, and mark in the asymptotes.

2.12 By first factorising the equation, or otherwise, sketch the function $(x^2 + y^2)^2 - 4x^2 = 0$.

3 Trigonometry

3.1 Angles and circular measure

An angle is a measure of rotation or turn, and is usually specified in *degrees*. There are 360° in a complete twist, where one ends up facing the same way as at the beginning; 90° therefore represents a right-angle, 180° an about-face, and so on. Despite our familiarity with degrees, a circular measure that is often more useful in a mathematical context is a *radian*. It is a dimensionless quantity defined as follows: if a radius of length R is spun through an angle θ and generates an arc of length L, then

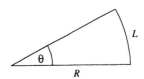

$$\theta = \frac{L}{R} \tag{3.1}$$

Since the circumference of a circle is $2\pi R$, $360° = 2\pi$ radians; a right-angle is $\pi/2$ and, in general, an angle in degrees can be converted into one in radians by multiplying it by $\pi/180$. In other words, a radian is about 57.3°.

Unless degrees are mentioned explicitly, it is best to assume that all angles are implicitly given in radians (especially if they contain factors of π).

3.2 Sines, cosines and tangents

The most elementary definition of sines, cosines, and tangents is provided by the ratios of the sides of a right-angled triangle. That is to say, if the length of the hypotenuse is r, and that of the sides adjacent and opposite to the angle θ is x and y respectively, then

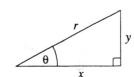

$$\sin\theta = \frac{y}{r}, \quad \cos\theta = \frac{x}{r}, \quad \tan\theta = \frac{y}{x} \tag{3.2}$$

From this, it is easy to see that the three trigonometric quantities are linked through the relationship

$$\tan\theta = \frac{\sin\theta}{\cos\theta} \tag{3.3}$$

Furthermore, as the angle between y and r is $90° - \theta$, because the sum around a triangle is 180°, the sines and cosines of (acute) angles are complementary

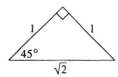

$$\sin\theta = \cos(\pi/2 - \theta) \tag{3.4}$$

If θ is extremely small, then y must be tiny while r and x are of virtually the same size; as such, eqns (3.2) and (3.4) tell us that $\sin(0) = \cos(\pi/2) = 0$ and $\cos(0) = \sin(\pi/2) = 1$. With some appropriate right-angled triangles, and Pythagoras' theorem, it can also be shown that $\sin(\pi/6) = \cos(\pi/3) = 1/2$, $\sin(\pi/4) = \cos(\pi/4) = 1/\sqrt{2}$, and $\sin(\pi/3) = \cos(\pi/6) = \sqrt{3}/2$. In conjunction

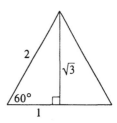

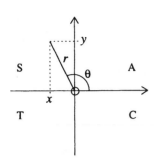

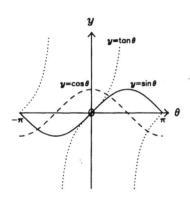

with eqn (3.3), therefore, we get $\tan(0) = 0$, $\tan(\pi/6) = 1/\sqrt{3}$, $\tan(\pi/4) = 1$, $\tan(\pi/3) = \sqrt{3}$, and $\tan(\pi/2) \to \infty$.

Our discussion has so far been restricted to the case of $0° \leq \theta \leq 90°$. It can be extended to any angle if we generalise our definitions so that x, y, and r in eqn (3.2) represent the coordinates, and distance from the origin, of a point whose angular displacement is θ when measured anticlockwise from the positive x-axis; a negative value of θ refers to a clockwise rotation. Then we find that the sine, cosine, and tangent of an arbitrary angle can always be related to one that lies between 0° and 90°, apart from a possible minus sign. To see this, the first thing to notice is that the trigonometric functions are *periodic* in that they identically repeat themselves every 360°

$$\sin\theta = \sin(\theta + 2\pi N) \qquad (3.5)$$

where N is an integer, and similarly for cos and tan. If $90° < \theta \leq 180°$, so that x is negative and y is positive, then $\tan\theta = y/x$ is equal to minus the 'opposite over the adjacent' of a right-angled triangle with the supplementary angle of $180° - \theta$; in other words, $\tan\theta = -\tan(\pi - \theta)$. The ratios of x and y with respect to the hypotenuse r yield $\sin\theta = \sin(\pi - \theta)$ and $\cos\theta = -\cos(\pi - \theta)$. A repetition of this argument for the cases $180° < \theta \leq 270°$ and $270° < \theta \leq 360°$ leads to the general result that the sine, cosine, and tangent of θ is equal to the value of the trigonometric function for the smallest corresponding angle between the 'radius' and the x-axis, but with a sign dependent on the quadrant. The *parity* of the latter can be remembered with the acronym CAST, which is made up from the first letters of 'cosine, all, sine, and tangent', as they are placed anticlockwise around the origin starting from the bottom right-hand corner, and refers to the functions that are positive in that quarter. For example, $\tan(210°) = \tan(30°)$, and $\tan(300°) = -\tan(60°)$.

Before leaving this section, we should make two further points. The first is simply a matter of nomenclature

$$\sec\theta = \frac{1}{\cos\theta}, \quad \mathrm{cosec}\,\theta = \frac{1}{\sin\theta}, \quad \cot\theta = \frac{1}{\tan\theta} \qquad (3.6)$$

where the abbreviations stand for secant, cosecant and cotangent. The second concerns the approximation of sin, cos, and tan when the angle is small but finite. We saw earlier that $\sin(0) = 0$, $\cos(0) = 1$, and $\tan(0) = 0$, but a more careful consideration of how the sector in the definition of circular measure starts to resemble a right-angled triangle as θ tends to zero leads to

$$\sin\theta \approx \theta, \quad \tan\theta \approx \theta, \quad \cos\theta \approx 1 - \theta^2/2 \qquad (3.7)$$

where $|\theta| \ll 1$, and is specified in radians. It can also be verified by looking at the plots of the trigonometric functions in the neighbourhood of $\theta = 0$.

3.3 Pythagorean identities

One of the most basic properties of a right-angled triangle that we learn about is Pythagoras' theorem; in our present set-up, it takes the form $x^2 + y^2 = r^2$. If we divided both sides by r^2, so that $(x/r)^2 + (y/r)^2 = 1$, then substitution from eqn (3.2) would yield the relationship

$$\sin^2\theta + \cos^2\theta = 1 \qquad (3.8)$$

where we've followed the convention that $\sin^2\theta = (\sin\theta)^2$, and so on. This formula, along with many others in this chapter, is often written with a three-pronged equals sign ($\equiv$) to indicate that it is an *identity*; this differs from an ordinary equation (like $\sin\theta = \cos\theta$) in that it holds for all values of θ, rather than just a few specific ones. Similar divisions of Pythagoras' theorem by x and y, instead of r, give rise to two additional identities

$$\tan^2\theta + 1 = \sec^2\theta \quad \text{and} \quad \cot^2\theta + 1 = \operatorname{cosec}^2\theta \qquad (3.9)$$

Incidentally, eqn (3.8) yields eqn (2.10) if $x = a\cos\theta$ and $y = b\sin\theta$; this is known as the *parametric* form of an ellipse. The equivalent formulation for a circle (centred at the origin) is $x = R\cos\theta$ and $y = R\sin\theta$.

3.4 Compound angles

If an angle is expressed as the sum of two others, such as $\theta = A + B$, then its sine can be written in terms of the trigonometric functions of its constituent parts. While the derivation of the formula requires a bit of geometrical and algebraic inspiration, the result is easy to state

$$\sin(A + B) = \sin A \cos B + \cos A \sin B \qquad (3.10)$$

A special case occurs when $A = B$ with the sine of a double-angle reducing to

$$\sin 2A = 2\sin A \cos A \qquad (3.11)$$

The sine of a difference between two angles, say $A - B$, can be obtained from eqn (3.10) by noting that $\sin(-\theta) = -\sin\theta$ and $\cos(-\theta) = \cos\theta$

$$\sin(A - B) = \sin A \cos B - \cos A \sin B \qquad (3.12)$$

The same geometrical construction, and similar algebraic reasoning, leads to the following formulae for the cosine of the sum and difference $A \pm B$

$$\cos(A \pm B) = \cos A \cos B \mp \sin A \sin B \qquad (3.13)$$

where the negative sign on the right-hand side goes with the positive one on the left, and vice versa. If $A = B$, then the cosine of a double-angle reduces to

$$\cos 2A = \cos^2 A - \sin^2 A \qquad (3.14)$$

With a substitution from eqn (3.8), it can also be written in two alternative forms that either contain only $\sin A$ or just $\cos A$

$$\cos 2A = 2\cos^2 A - 1 = 1 - 2\sin^2 A \qquad (3.15)$$

According to eqn (3.3), the divisions of eqns (3.10) and (3.12) by their counterparts in eqn (3.13) give the formulae for the tangent of the sum and difference of two angles

$$\tan(A \pm B) = \frac{\tan A \pm \tan B}{1 \mp \tan A \tan B} \qquad (3.16)$$

The result $\tan 2A = 2\tan A/(1 - \tan^2 A)$ then follows from putting $A = B$.

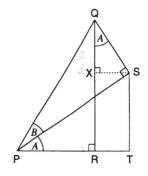

$$\sin(A + B) = \frac{QR}{PQ}$$

$$= \frac{QX + ST}{PQ}$$

$$= \frac{QS \cos A + PS \sin A}{PQ}$$

$$= \frac{QS}{PQ}\cos A + \frac{PS}{PQ}\sin A$$

$$= \sin B \cos A + \cos B \sin A$$

3.5 Factor formulae

Useful identities, relating the sums and differences of sines and cosines to their products, can be derived from the formulae of the previous section. For example, the addition of eqns (3.10) and (3.12) gives

$$\sin(A+B) + \sin(A-B) = 2\sin A \cos B \tag{3.17}$$

The simple substitution $X = A + B$ and $Y = A - B$ then yields

$$\sin X + \sin Y = 2\sin\left(\frac{X+Y}{2}\right)\cos\left(\frac{X-Y}{2}\right) \tag{3.18}$$

$$\sin X - \sin Y =$$

$$2\cos\left(\frac{X+Y}{2}\right)\sin\left(\frac{X-Y}{2}\right)$$

Although these equations are equivalent, the former is better for decomposing a product into a sum and the latter is more suitable for going the other way. The subtraction of eqns (3.10) and (3.12) leads to expressions similar to eqns (3.17) and (3.18), except that there is a minus sign on the left-hand side and an interchange between the sin and cos on the right.

A corresponding manipulation for the cosines of compound angles, from eqn (3.13), leads to

$$\cos X - \cos Y =$$

$$-2\sin\left(\frac{X+Y}{2}\right)\sin\left(\frac{X-Y}{2}\right)$$

$$\cos(A+B) + \cos(A-B) = 2\cos A \cos B \tag{3.19}$$

and

$$\cos(A+B) - \cos(A-B) = -2\sin A \sin B \tag{3.20}$$

along with their X and Y counterparts.

3.6 Inverse trigonometric functions

If we knew that the sine of an angle θ in a right-angled triangle was equal to a half, then we could deduce that θ was 30°. This kind of reverse operation is an example of using an *inverse* trigonometric function. That is to say, the case above could be generalised with the formal definition

$$y = \sin\theta \quad\Longleftrightarrow\quad \theta = \sin^{-1}y \tag{3.21}$$

We should caution that the 'power of minus one' notation here is anomalous, though widespread, in that it does not mean the reciprocal of $\sin y$; a less confusing, but rarer, alternative for the inverse function is arcsin. Expressions very similar to eqn (3.21) can be written for all the trigonometric functions, where $\arccos y = \cos^{-1}y$, $\arctan y = \tan^{-1}y$, and so on.

$$\cos^{-1}(1/2) = \pm\tfrac{\pi}{3}, \pm\tfrac{7\pi}{3}, \ldots$$

$$\tan^{-1}\sqrt{3} = \tfrac{\pi}{3}, \tfrac{\pi}{3} \pm \pi, \ldots$$

To appreciate the nature of the relationship $\theta = \sin^{-1}y$ we simply need to turn the plot of $y = \sin\theta$ through 90°, so that the y-axis is horizontal and the θ-axis is vertical; the same is true for $\theta = \cos^{-1}y$ and $\theta = \tan^{-1}y$ with respect to the graphs of cos and tan. It then follows that the inverse trigonometric functions are *multivalued*, since a given sin, cos, or tan can be obtained from more than one angle. If we were only told that $\sin\theta = 1/2$, for example, then θ could be 30° or 150°; or, indeed, any multiple number of 360° in addition to these two values. The other main point to note is that $\sin^{-1}y$ and $\cos^{-1}y$ do not exist if the magnitude of y is greater than unity, because $|\sin\theta| \leq 1$ and $|\cos\theta| \leq 1$ for all θ; there is no such restriction for $\tan^{-1}y$, of course, as $\tan\theta$ can lie anywhere in the range $\pm\infty$.

3.7   The sine and cosine rules

Let's conclude this chapter with a statement of the sine and cosine formulae, which relate the lengths of the sides of a general triangle to its angles; their proofs will become straightforward after dealing with the topic of *vectors*. If the angles of a triangle are labelled as A, B, and C, then the convention is to denote the sides opposite them with the corresponding lowercase letters a, b, and c. With this set-up, it can be shown that

$$\frac{a}{\sin A} = \frac{b}{\sin B} = \frac{c}{\sin C} \qquad (3.22)$$

and

$$a^2 = b^2 + c^2 - 2bc \cos A \qquad (3.23)$$

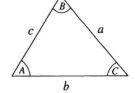

These are known as the sine and cosine rules , respectively. The former can be split up into three separate relationships, since this is the number of equal pairs in eqn (3.22). The latter also yields three equations, because the sides and angles in eqn (3.23) can be permuted; e.g. $b^2 = c^2 + a^2 - 2ca \cos B$.

Exercises

3.1 Convert the following angles from radians to degrees, and sketch their directions graphically: $\theta = \pi/6, \pi/4, \pi/3, \pm\pi/2, 2\pi/5, \pm 2\pi/3, \pi, 3\pi/2$.

3.2 Given the values of sin, cos and tan for $\pi/6$, $\pi/4$, and $\pi/3$, evaluate these trigonometric functions for $\pm 2\pi/3, \pm 3\pi/4, \pm 5\pi/6, 5\pi/4, 4\pi/3, -\pi/6$.

3.3 From the definitions of a radian and a sine, indicate why $\sin\theta \approx \theta$ for small angles; show how eqn (3.15) then leads to $\cos\theta \approx 1 - \theta^2/2$.

3.4 If $t = \tan(\theta/2)$, express $\tan\theta$, $\cos\theta$ and $\sin\theta$ in terms of t.

3.5 Solve the following in the range $-\pi$ to π: $\tan\theta = -\sqrt{3}$, $\sin 3\theta = -1$, and $4\cos^3\theta = \cos\theta$.

3.6 Show that $a\sin\theta + b\cos\theta$ can be written as $A\sin(\theta + \phi)$, where A and ϕ are related to a and b. Hence solve the equation $\sin\theta + \cos\theta = \sqrt{3/2}$.

3.7 Show that $\cos 4\theta = 8\cos^4\theta - 8\cos^2\theta + 1$, and express $\sin 4\theta$ in terms of $\sin\theta$ and $\cos\theta$.

3.8 Show that $8\sin^4\theta = \cos 4\theta - 4\cos 2\theta + 3$, and find a similar expression for $\cos^4\theta$.

3.9 By using a factor formula, find the values of θ between 0 and π which satisfy the equation $\cos\theta = \cos 2\theta + \cos 4\theta$.

3.10 Show that $\cos\theta + \cos 3\theta + \cos 5\theta + \cos 7\theta = 4\cos\theta \cos 2\theta \cos 4\theta$.

3.11 A triatomic molecule has bond-lengths of 1.327Å and 1.514Å, and a bond-angle of 107.5°; find the distance between the two furthest atoms.

4 Differentiation

4.1 Gradients and derivatives

In chapter 2 we saw how the relationship between two quantities, called x and y, could be visualised with the aid of a graph. While the intersections of the curve with the x and y axes may be of interest, it is often more important to know the slope at any given point; that is, how quickly y increases, or decreases, as x changes, and vice verse. This issue is at the heart of the topic of *differentiation*, and most of this chapter is devoted to learning how to calculate the gradient algebraically.

Let us begin with a precise definition of what is meant by the slope of a curve. Suppose that y is related to x through some function called 'f', usually written as $y = f(x)$, so that $f(x) = mx + c$ for a general straight line, and $f(x) = \sin(x)$ for a sinusoidal variation, and so on. Then, if the horizontal coordinate changes from x to $x + \delta x$, where δx represents a small increment, the value of y is altered from $f(x)$ to $f(x + \delta x)$. The gradient, at a point x, is defined to be the ratio of the change in the vertical coordinate, δy, to that of the horizontal increment, as δx becomes infinitesimally small. This can be stated formally as

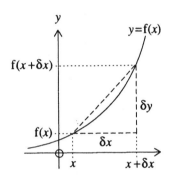

$$\frac{dy}{dx} = \lim_{\delta x \to 0} \left[\frac{\delta y}{\delta x} \right] = \lim_{\delta x \to 0} \left[\frac{f(x + \delta x) - f(x)}{\delta x} \right] \tag{4.1}$$

where dy/dx is known as the *derivative*, or *differential coefficient*, and is pronounced 'dy-by-dx'; it can also be denoted with a 'dashed' notation, y' or $f'(x)$, or even a 'dotted' one, $\dot{y}$, if the x-axis pertains to time. The tendency of $\delta x \to 0$ has to be approached gradually to ascertain the *limiting* value of the ratio $\delta y / \delta x$, as both increments are individually equal to zero when the condition is met. Strictly speaking, we should check that the same value of dy/dx is obtained whether δx is positive or negative, but this is assured as long as the curve $y = f(x)$ is 'smooth'; inconsistencies will arise if kinks and sudden breaks (or discontinuities) are present, and the function is said to be non-differentiable at those points.

As a very simple example of evaluating a derivative from 'first principles', let's consider the case of a straight line $y = mx + c$. Substituting for $f(x)$ into eqn (4.1) gives

$$\frac{dy}{dx} = \lim_{\delta x \to 0} \left[\frac{m(x + \delta x) + c - (mx + c)}{\delta x} \right] = \lim_{\delta x \to 0} \left[\frac{m\,\delta x}{\delta x} \right]$$

where the last term follows from elementary algebra. Although the numerator and denominator both become zero in the limit $\delta x \to 0$, their ratio is well-defined; indeed, as expected from the discussion in section 2.1, the derivative

$$y' = \frac{dy}{dx} = \frac{df}{dx} = f'(x)$$

$$\dot{y} = \frac{dy}{dt}$$

has the same value everywhere and is equal to the gradient m. Although we have specifically dealt with a straight line, there are a couple of points that apply more generally. The first is that y increases with x if $dy/dx>0$, and it decreases if $dy/dx<0$; the rate of the variation is given by the magnitude of the derivative. The second point is that

$$\frac{dy}{dx} = \tan\theta \qquad (4.2)$$

where θ is the angle between the slope and the x-axis. The properties of a tangent also confirm the former observation, since dy/dx has the same sign as θ and a magnitude which increases with the size of the angle.

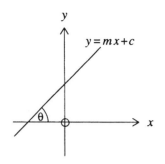

Having illustrated differentiation with an elementary example, let's move on to a more interesting case: $y = \sin x$. The substitution of $f(x)$ into eqn (4.1) now gives

$$\frac{dy}{dx} = \lim_{\delta x \to 0} \left[\frac{\sin(x+\delta x) - \sin x}{\delta x} \right]$$

As always, the taking of the limit will be the very last step; first we must consider how the numerator behaves as δx becomes tiny, but isn't actually zero. This can be done by expanding $\sin(x+\delta x)$ with eqn (3.10), and using the small-angle approximations for sine and cosine in eqn (3.7)

$$\sin(x+\delta x) \approx \left(1 - \frac{\delta x^2}{2}\right) \sin x + \delta x \cos x$$

where x is implicitly measured in radians. Putting this into our definition for dy/dx, and simplifying the resultant expression, we obtain

$$\frac{dy}{dx} = \lim_{\delta x \to 0} \left[\cos x - \frac{\delta x \sin x}{2} \right]$$

Hence on finally letting δx go to zero, we see that the derivative of $\sin x$ is $\cos x$. This can be verified from the plots of sine and cosine in section 3.2, although the x-axis there was labelled as θ. At the origin, the slope of the $\sin x$ curve is inclined at 45°; its derivative is therefore $\tan(\pi/4) = 1$, which is the same as $\cos(0)$. The gradient reduces gradually to zero between 0° and 90°, as does the cosine. The sine curve then starts to dip downwards so that, just like $\cos x$, the slope becomes increasingly negative. And so on, for all values of x. A similar algebraic argument leads to the result that the derivative of $\cos x$ is $-\sin x$, and can also be verified graphically.

$$\frac{d}{dx}(\sin x) = \cos x$$

$$\frac{d}{dx}(\cos x) = -\sin x$$

4.2 Some basic properties of derivatives

One of the most fundamental properties of derivatives is their *linearity*. That is to say, they satisfy the rules

$$\frac{d}{dx}[A f(x)] = A \frac{df}{dx} \quad \text{and} \quad \frac{d}{dx}[f(x) + g(x)] = \frac{df}{dx} + \frac{dg}{dx} \qquad (4.3)$$

where A is a constant, $f(x)$ and $g(x)$ are arbitrary functions of x, and d/dx is the differential *operator* meaning 'the rate of change, with respect to x, of'. While the formal proof of eqn (4.3) follows from eqn (4.1), the essential point is that 'the derivative of a sum is equal to the sum of the derivatives'. This enables us to differentiate a polynomial, for example, by knowing that the derivative of x^M is Mx^{M-1}

$$\frac{d}{dx}\left(a_0 + a_1 x + a_2 x^2 + \cdots + a_N x^N\right) = a_1 + 2a_2 x + \cdots + Na_N x^{N-1} \qquad (4.4)$$

$$\frac{d}{dx}\left(x^M\right) = Mx^{M-1}$$

The differential coefficient of x^M itself can be derived from eqn (4.1) by using the binomial expansion of $(x + \delta x)^M$, as in eqn (1.11), and noting that all the terms with powers of δx greater than one are negligible. Although the result is only shown for positive integer values of M in this way, it turns out to hold for any power. Eqn (4.4) also confirms our earlier finding that the derivative of a straight line, which corresponds to $N = 1$, is a constant (a_1); as expected, the gradient of the case with no variation, $y = a_0$, is zero.

Another important property of dy/dx is its reciprocal relationship with its counterpart dx/dy

$$\frac{dy}{dx} = \frac{1}{dx/dy} \qquad (4.5)$$

While the former is the slope of a curve in an ordinary graph, the latter is the equivalent gradient when y is plotted horizontally and x vertically; in other words, dx/dy is the rate of change of x with respect to y. The proof of eqn (4.5) is almost self-evident, in that it's obviously true for arbitrarily small (but finite) increments δx and δy.

A good example of the use of eqn (4.5) is provided by the derivatives of inverse trigonometric functions. Following eqn (3.21), we have

$$y = \sin^{-1} x \qquad \Longleftrightarrow \qquad x = \sin y$$

The differentiation of the right-hand expression with respect to y gives

$$\frac{dx}{dy} = \cos y = \sqrt{1 - \sin^2 y} = \sqrt{1 - x^2}$$

$$\frac{d}{dx}\left(\sin^{-1} x\right) = \frac{1}{\sqrt{1 - x^2}}$$

$$\frac{d}{dx}\left(\cos^{-1} x\right) = \frac{-1}{\sqrt{1 - x^2}}$$

where we have used the identity of eqn (3.8) to relate $\cos y$ to $\sin y$, and hence write dx/dy in terms of x; we have implicitly assumed that $|x| \leq 1$, and $|y| \leq \pi/2$. According to eqn (4.5), therefore, the derivative of $\arcsin x$ is equal to the reciprocal of the square root of $1 - x^2$. A similar analysis for $\arccos x$ shows that its derivative is minus that for $\arcsin x$.

The final point in this section is about the differentiation of derivatives. If, like $y = f(x)$, $dy/dx = f'(x)$ is a 'smooth' function of x, then it too can be differentiated to yield the *second derivative*

$$y'' = \frac{d^2 y}{dx^2} = \frac{d}{dx}\left(\frac{dy}{dx}\right) = \lim_{\delta x \to 0}\left[\frac{f'(x + \delta x) - f'(x)}{\delta x}\right] = f''(x) \qquad (4.6)$$

$$\frac{d^n y}{dx^n} = \frac{d}{dx}\left(\frac{d^{n-1} y}{dx^{n-1}}\right)$$

Rather than telling us how y changes with x, it conveys how the slope of y varies with x. If y represents *distance* travelled, and x time, for example, then y' (or $\dot{y}$) gives the *speed* and y'' (or $\ddot{y}$) the *acceleration*. Eqn (4.6) generalises to higher order derivatives, so that the third one, $d^3 y/dx^3$, is the rate of change of $d^2 y/dx^2$ with respect to x ; and so on.

4.3 Exponentials and logarithms

In section 2.4, we encountered the exponential function $y = \exp(x)$. This has the property that its gradient varies with x in the same way as the function itself

$$\frac{d}{dx}(e^x) = e^x \qquad (4.7)$$

This can even be regarded as the definition of the number 'e'. The derivative of the natural logarithm, $y = \ln(x)$, then follows from the differentiation of the equivalent exponential expression, $x = \exp(y)$, with respect to y

$$\frac{dx}{dy} = e^y = x$$

Hence, with the reciprocal relationship of eqn (4.5), we have

$$\frac{d}{dx}[\ln(x)] = \frac{1}{x} \qquad (4.8)$$

The derivative of a logarithm, or a power, to any other base, say a, can be obtained by using the results of section 1.2 in conjunction with those above. For example, eqn (1.8) allows us to write $\log_a(x) = \ln(x)/\ln(a)$; this means that the derivative of $\log_a(x)$ is equal to that of $\ln(x)$, but divided by $\ln(a)$. Similarly, the derivative of a^x turns out to be just a^x times $\ln(a)$.

$$e = 2.7182818285$$

$$\frac{d}{dx}(a^x) = a^x \ln(a)$$

$$\frac{d}{dx}[\log_a(x)] = \frac{1}{x \ln(a)}$$

4.4 Products and quotients

The linearity of eqn (4.3) tells us how to evaluate the differential coefficient of a sum, $y = u(x) + v(x)$, where u and v are arbitrary functions of x, but how do we calculate it for a product $y = u(x)v(x)$? The answer can be derived readily from eqn (4.1), as long as we note that $u(x + \delta x)$ and $v(x + \delta x)$ may be expressed as $u + \delta u$ and $v + \delta v$ respectively, where δu and δv are the small changes in the values of u and v generated by the increment δx. Then, by the definition of a derivative, we have

$$\frac{dy}{dx} = \lim_{\delta x \to 0}\left[\frac{(u + \delta u)(v + \delta v) - uv}{\delta x}\right] = \lim_{\delta x \to 0}\left[u\frac{\delta v}{\delta x} + v\frac{\delta u}{\delta x} + \frac{\delta u\,\delta v}{\delta x}\right]$$

and on taking the limit $\delta x \to 0$, we obtain

$$\frac{d}{dx}(uv) = u\frac{dv}{dx} + v\frac{du}{dx} \qquad (4.9)$$

because $\delta u\,\delta v/\delta x \to 0$. This formula can be extended to the product of any number of terms, by either returning to first principles or by putting $v = fg$.

$$(uvw)' = uvw' + uv'w + u'vw$$

As an example of the use of eqn (4.9), consider the function $y = 2^x(x^3 - 1)$. This is equivalent to having $u = 2^x$ and $v = x^3 - 1$, so that their derivatives are $u' = 2^x \ln(2)$ and $v' = 3x^2$, whereby

$$\frac{d}{dx}[2^x(x^3 - 1)] = 2^x[3x^2 + (x^3 - 1)\ln(2)]$$

With practice, the substitution of the functions u and v into eqn (4.9) is not required; rather, it's simpler to implement it directly by remembering that the differential coefficient of a product is given by 'the first one times the derivative of the second, plus the second times the derivative of the first'.

$$(uv)'' = uv'' + 2u'v' + u''v$$

Incidentally, the repeated differentiation of a product of two terms leads to an expression analogous to the binomial expansion of eqn (1.11)

$$\frac{d^n}{dx^n}(uv) = \sum_{r=0}^{n} {}^nC_r \frac{d^r u}{dx^r} \frac{d^{n-r} v}{dx^{n-r}} \tag{4.10}$$

This is known as Leibnitz's theorem.

The derivative of a *quotient* or ratio, $y = u(x)/v(x)$, may be ascertained by applying the product rule to $u = vy$; the resulting equation, $u' = vy' + yv'$, can then be manipulated algebraically to yield y'

$$\frac{d}{dx}\left(\frac{u}{v}\right) = \frac{vu' - uv'}{v^2} \tag{4.11}$$

A straightforward example of the use of this formula is in the calculation of the differential coefficient of $\tan x$:

$$\frac{d}{dx}(\tan x) = \frac{d}{dx}\left(\frac{\sin x}{\cos x}\right) = \frac{\cos^2 x + \sin^2 x}{\cos^2 x}$$

$$\frac{d}{dx}(\tan x) = \sec^2 x$$

$$\frac{d}{dx}(\tan^{-1} x) = \frac{1}{1+x^2}$$

$$\frac{d}{dx}(\cot x) = -\operatorname{cosec}^2 x$$

where we have assumed a knowledge of the derivatives of $\sin x$ and $\cos x$, as well as eqns (3.3) and (4.11). Since the numerator on the far right is unity, by eqn (3.8), the differential coefficient of $\tan x$ is equal to one over $\cos^2 x$; or, with eqn (3.6), just $\sec^2 x$. This analysis can be retraced to show that the derivative of $\cot x$ is $-\operatorname{cosec}^2 x$, or the result for the tangent used directly to differentiate its inverse form $y = \tan^{-1} x$ (as for the arcsin earlier); the latter turns out to be given by the reciprocal of $1 + x^2$.

4.5 Functions of functions

We have now seen how to differentiate power-laws, exponentials, logarithms, trigonometric functions, and arithmetical combinations of them. Frequently, however, we encounter these familiar entities in more complicated settings; for example, $y = \ln(2 + \cos x)$, or $y = A \sin^2(\omega x + \phi)\exp(-kx)$ where all but x and y are constant. How can we use our knowledge about the derivatives of the underlying 'building blocks' to deal with the general situation?

Well, we need to enlist the help of a very powerful result known as the *chain rule*. This states that if y is a function of u, and u is itself dependent on x, then dy/dx is given by

$$\frac{dy}{dx} = \frac{dy}{du} \times \frac{du}{dx} \tag{4.12}$$

$$\frac{d}{dx}(\sec x) = \sec x \tan x$$

$$\frac{d}{dx}(\operatorname{cosec} x) = -\operatorname{cosec} x \cot x$$

Thus in our first example, $y = \ln(u)$ and $u = 2 + \cos x$; since $dy/du = 1/u$ and $du/dx = -\sin x$, eqn (4.12) yields $dy/dx = -\sin x/(2 + \cos x)$. Several other straightforward cases, such as $y = \sec x$ and $y = \operatorname{cosec} x$, fall under the reciprocal category of $y = u^{-1}$; their derivatives, therefore, share the common form $dy/dx = -u^{-2}du/dx = -u'/u^2$. Indeed, the quotient rule of eqn (4.11) can be derived from eqn (4.9) by considering u/v to be the product of u and $1/v$.

The proof of eqn (4.12) is, again, almost self-evident, in that it holds for arbitrarily small increments δx, δy, and δu through simple division. In fact this indicates that the chain rule can be extended, as required, to deal with nested sets of functions. If $y = \sin[\ln(2 + \cos x)]$, for example, then we have $y = \sin u$, where $u = \ln(v)$ and $v = 2 + \cos x$; thus, by putting the 'standard' derivatives, like $dy/du = \cos u$, into

$$\frac{dy}{dx} = \frac{dy}{du} \times \frac{du}{dv} \times \frac{dv}{dx}$$

we obtain $dy/dx = -\cos[\ln(2 + \cos x)] \sin x/(2 + \cos x)$. With experience, the explicit substitution of u, v, ..., becomes unnecessary; familiarity eventually allows the chain rule to be implemented mentally in a stepwise manner.

Several of the differentiation rules that we have learnt may be needed for any given problem, and sometimes more than once. If $y = x\ln[x(2 + \cos x)]$, for example, then we have to deal with the product of x and $\ln(u)$, where u is itself a product of x and $2 + \cos x$. Thus $dy/dx = xu^{-1}du/dx + \ln u$, where $du/dx = -x\sin x + 2 + \cos x$.

As a final note in this section, we should mention an important variant of the chain rule

$$\frac{dy}{dx} = \frac{dy/dt}{dx/dt} \tag{4.13}$$

which is useful for differentiating parametric equations. That is when both x and y are expressed as functions of a common variable, denoted by t in eqn (4.13). An example was met in section 3.3, where an ellipse was seen to take the form $x = a\cos\theta$ and $y = b\sin\theta$; it follows that $dy/dx = -b\cot\theta/a$.

4.6 Maxima and minima

One of the main reasons for being interested in the slopes of curves is that the places where the gradient is zero are often of great physical significance. They can be found by solving the equation

$$\frac{dy}{dx} = 0 \tag{4.14}$$

and are known as *stationary* points. As the name suggests, they are locations where the value of y remains virtually unchanged even if x varies a little bit; in other words, they represent positions of 'equilibrium'.

Apart from the cases where y gradually approaches a constant value as x tends to infinity, as in $y = \exp(-x)$, there are three types of situations when eqn (4.14) is satisfied: (i) a *maximum*, which is like the top of a hill, where y decreases on both sides; (ii) a *minimum*, which is akin to the base of a valley, so that y increases around the stationary point; and (iii) a point of *inflexion*, which is a flat region where y goes up on one side and down on the other. The first two scenarios are collectively called 'turning points', and can be distinguished by considering how the sign of the gradient dy/dx changes as the slope passes through the horizontal position

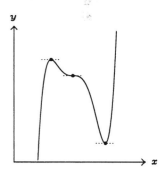

$$\frac{d^2y}{dx^2} < 0 \quad \text{for a max,} \quad \text{and} \quad \frac{d^2y}{dx^2} > 0 \quad \text{for a min} \tag{4.15}$$

Although the second derivative is zero at a point of inflexion, the condition that $d^2y/dx^2 = 0$ doesn't necessarily imply one. Indeed, this is a special case that always requires more careful thought. For example, it is easily shown that the curves $y = x^3$ and $y = x^4$ both have $dy/dx = 0$ and $d^2y/dx^2 = 0$ at the origin; from

$$y = x^4$$
$$y' = 4x^3$$
$$y'' = 12x^2$$

the graphs in section 2.3, however, it's clear that the quartic has a minimum while the cubic harbours a point of inflexion.

As a concrete illustration of the discussion above, let's consider a hydrogen atom. *Quantum mechanics* shows that the probability that an electron in the 1s (ground) state is at a distance r from the nucleus is proportional to

$$p(r) = \frac{4r^2}{a_0^3} e^{-2r/a_0}$$

where the constant $a_0 = 0.5292 \times 10^{-10}$ m. The most likely radial position of the electron is given by the maximum of $p(r)$, therefore, which is defined by the criterion of eqn (4.14)

$$\frac{dp}{dr} = \frac{8r}{a_0^3} \left(1 - \frac{r}{a_0}\right) e^{-2r/a_0} = 0$$

Of the three possible solutions, $r = 0$, $r = a_0$, and $r \to \infty$, only the middle one yields a negative value for the second derivative d^2p/dr^2; hence, according to eqn (4.15), the most probable location of the electron is at $r = a_0$.

There are many problems of physical interest where the minimum value of a function, such as the *Gibbs free energy*, or the mismatch of a model fit to experimental data, is of paramount importance. An example familiar from everyday experience is that of gravitational *potential energy*, where objects, be they marbles or bodies of water, settle naturally into the nearest hollow; as such, a minimum is known as a point of 'stable' equilibrium. By contrast, a maximum gives an 'unstable' equilibrium.

4.7 Implicit and logarithmic differentiation

So far, we have generally assumed that y is an explicit function of x; often, however, the relationship between the two entities is sufficiently complicated that a rearrangement into the form $y = f(x)$ is not possible. Nevertheless, dy/dx can still be ascertained, in terms of x and y, by applying the operator d/dx to both sides of the equation. If $x^2y + \sin y = 6$, for example, then

$$\frac{d}{dx}(x^2y) + \frac{d}{dx}(\sin y) = \frac{d}{dx}(6)$$

$$\frac{d}{dx} = \frac{dy}{dx} \times \frac{d}{dy}$$

where we have used the linearity property of eqn (4.3) to split up the left-hand side into two parts. By the rules of differentiation for products, functions of functions, and so on, we therefore obtain

$$x^2 \frac{dy}{dx} + 2xy + \cos y \frac{dy}{dx} = 0$$

which can be manipulated algebraically to yield $dy/dx = -2xy/(x^2 + \cos y)$. This procedure, where an expression is differentiated 'as it stands', is usually called *implicit differentiation*.

Even when y can be written as an explicit function of x, it is sometimes helpful to take its logarithm before differentiating it. This is particularly true for cases involving complicated products and quotients, such as

$$y = \frac{(x+5)\sqrt{(7+2x)^3}}{(2x^3+1)\cos x}$$

because, according to the rules of section 1.2, it turns the expression into a simpler one of sums and differences

$$\ln(y) = \ln(x+5) + 3\ln(7+2x)/2 - \ln(2x^3+1) - \ln(\cos x)$$

which is considerably easier to differentiate (implicitly), term by term. This logarithmic technique is also very useful for dealing with situations where x occurs as a power. For example, if $y = (x^2+1)^x$ then an application of the differential operator d/dx to both sides of $\ln(y) = x \ln(x^2+1)$, and a little algebra, yields $dy/dx = y[\ln(x^2+1) + 2x^2/(x^2+1)]$.

$$\frac{d}{dx}[\ln(y)] = \frac{dy}{dx} \times \frac{1}{y}$$

Exercises

4.1 From 'first principles', differentiate: (i) $y = \cos x$; (ii) $y = x^n$, where n is a positive integer; (iii) $y = 1/x$; and (iv) $y = 1/x^2$.

4.2 For which values of x are the following not differentiable: (i) $y = |x|$, (ii) $y = \tan x$, and (iii) $y = e^{-|x|}$.

4.3 Give an argument for why the gradient of a straight line perpendicular to $y = mx + c$ is $-1/m$.

4.4 By exploiting the linearity property of a differential operator, and using a knowledge of the sum of an infinite GP, show that

$$\sum_{n=1}^{\infty} n x^{n-1} = \frac{d}{dx}\left(\frac{x}{1-x}\right)$$

where $|x| < 1$. Hence, evaluate the summation.

4.5 Differentiate $y = \cos^{-1}(x/a)$, where $|x/a| < 1$; is your result valid for all values of y? What is the derivative of $\tan^{-1}(x/a)$?

4.6 Differentiate the following with respect to x: (i) $(2x+1)^3$; (ii) $\sqrt{3x-1}$; (iii) $\cos 5x$; (iv) $\sin(3x^2+7)$; (v) $\tan^4(2x+3)$; (vi) $x \exp(-3x^2)$; (vii) $x \ln(x^2+1)$; (viii) $\sin x/x$.

4.7 By first taking logarithms, differentiate $y = a^x$ where a is a constant.

4.8 Find dy/dx when: (i) $x = t(t^2+2)$ and $y = t^2$; (ii) $x^2 = y \sin(xy)$.

4.9 Find and classify the stationary points of the following functions of x and r, where ε, σ and a_0 are constants:

(i) $f(x) = \dfrac{x^5}{5} - \dfrac{x^4}{6} - x^3$

(ii) $f(x) = \dfrac{x}{1+x^2}$

(iii) $U(r) = 4\varepsilon\left[\left(\dfrac{\sigma}{r}\right)^{12} - \left(\dfrac{\sigma}{r}\right)^{6}\right]$

(iv) $p(r) = \dfrac{r^2}{8a_0^3}\left(2 - \dfrac{r}{a_0}\right)^2 e^{-r/a_0}$

5 Integration

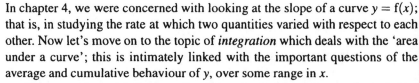

5.1 Areas and integrals

In chapter 4, we were concerned with looking at the slope of a curve $y = f(x)$; that is, in studying the rate at which two quantities varied with respect to each other. Now let's move on to the topic of *integration* which deals with the 'area under a curve'; this is intimately linked with the important questions of the average and cumulative behaviour of y, over some range in x.

To set up a formal definition of an integral, consider the region bounded by the straight lines $x = a$, $x = b$, and $y = 0$, and the curve $y = f(x)$. The size of the enclosure can be estimated by approximating it as a whole series of narrow vertical strips, and adding together the areas of these contiguous rectangular blocks. If the x-axis between a and b is divided into N equal intervals, then the width of each strip is given by $\delta x = (b - a)/N$; the corresponding heights of the thin blocks are equal to the values of the function $f(x)$ at their central positions. In other words, the area of the jth strip, which is at $x = x_j$ and of height $y = f(x_j)$, is $f(x_j)\,\delta x$; the index j ranges from 1 to N, of course, with $x_1 = a + \delta x/2$ and $x_N = b - \delta x/2$. As N tends to infinity, $\delta x \to 0$ and the approximation to the area under the curve becomes ever more accurate. This limiting form of the summation procedure defines an integral

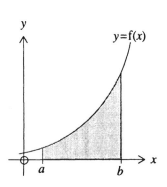

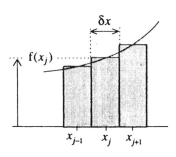

$$\int_a^b y\,dx = \int_a^b f(x)\,dx = \lim_{N\to\infty} \sum_{j=1}^N f(x_j)\,\delta x \qquad (5.1)$$

where the symbol $\int dx$ is read as the 'integral, from a to b, with respect to x'. The use of the term 'area' in the above discussion needs some qualification, in that it can be negative; this is because the 'height' of a strip $f(x_j) < 0$ whenever the curve $y = f(x)$ lies below the x-axis (and even the 'width' $\delta x < 0$ if $b < a$).

Before going on to see how integrals are evaluated, we should try to get a better feel for what they represent physically. As a simple example, suppose that a car moved in a straight line with a constant speed v_0 for a time t_0; then, it will have travelled a distance $v_0 t_0$. If the journey had progressed with a varying speed $v(t)$, as a function of time, how far would the car have gone? To work this out, the duration of the trip can be divided into a series of short intervals δt over which the speed is roughly constant. The distance travelled is then given by the sum of all the small contributions $v(t)\delta t$; that is to say, the integral of $v(t)$, with respect to t, from $t = 0$ to $t = t_0$

$$\text{Distance travelled } = \int_0^{t_0} v(t)\,dt \qquad (5.2)$$

Negative values of $v(t)$ can be understood as a backwards motion of the car, so that we should really talk about *velocity* rather than speed, and gives rise to a reduction in the (forward) length of the journey. The central rôle of an integral in the calculation of an average can also be appreciated, as the mean speed is the ratio of the overall distance travelled to the total time taken; in other words, the constant speed, $\langle v \rangle$ or $\bar{v}$, required for an equivalent trip is given by eqn (5.2) divided by t_0.

$$\bar{y} = \langle y \rangle = \frac{1}{b-a} \int_a^b y \, dx$$

5.2 Integrals and derivatives

Although an integral is defined as the limiting form of a summation, it is rarely ascertained from eqn (5.1). Rather, it's usually evaluated by noting that 'integration is the reverse of differentiation'. While this may not be obvious, it is easily confirmed for our earlier kinematic case: the distance travelled was seen to be the integral of the speed, and speed is the rate-of-change of distance (a derivative). To see this relationship more generally, consider the integral of $y = f(x)$ between the origin and an arbitrary point on the x-axis

$$\int_0^x y \, dx = G(x) \tag{5.3}$$

We have set this equal to a function $G(x)$ because the area under the curve will vary smoothly with the position of the upper bound. In particular, a small increment δx will add a contribution of approximately $f(x)\delta x$

$$G(x + \delta x) \approx G(x) + f(x)\delta x$$

This expression becomes exact in the limit $\delta x \to 0$, where upon an algebraic rearrangement, and eqn (4.1), shows that

$$f(x) = \lim_{\delta x \to 0} \left[\frac{G(x + \delta x) - G(x)}{\delta x} \right] = \frac{dG}{dx} \tag{5.4}$$

Hence, $G(x)$ is the function whose differential coefficient is equal to $f(x)$. Since the derivative of $\sin x$ is $\cos x$, for example, the integral of $\cos x$ is $\sin x$. Having obtained $G(x)$ in this manner, a simple consideration of the areas that $G(a)$ and $G(b)$ represent allows us to write the integral of eqn (5.1) as

$$\int_a^b f(x) \, dx = G(b) - G(a) = \left[G(x) \right]_a^b \tag{5.5}$$

where the square-bracket notation on the far right is a standard abbreviation for the difference of terms in the middle.

The calculation of most integrals hinges on eqns (5.4) and (5.5). That is to say: (i) thinking about differentiation backwards, to obtain $G(x)$; and (ii) the substitution of the limits a and b. While the second step is straightforward, the former often involves some preliminary manipulations; we will come to these shortly. An important general point, however, is that eqn (5.4) only determines $G(x)$ to within an arbitrary additive constant C (say); this is because $G(x)$ and $G(x) + C$ have the same differential coefficient since $dC/dx = 0$. This degree of uncertainty is of no consequence for eqn (5.5), as C cancels out on taking

$$\int_1^2 x \, dx = \left[\frac{x^2}{2} \right]_1^2$$
$$= 2 - \frac{1}{2} = \frac{3}{2}$$

$$G(x) = \int \sin x \, dx = C - \cos x$$

If $G(0) = 0$, then $C = 1$

the difference, but it does matter if the limits of the integration are not specified. The latter case is called an *indefinite* integral, in contrast to the *definite* ones of eqns (5.1) and (5.5), and requires an additional constraint, or *boundary condition*, such as the value of $G(0)$, to pin down the 'constant of integration'.

5.3   Some basic properties of integrals

Just as for the case of derivatives in section 4.2, one of the most fundamental properties of integrals is their linearity. That is to say, they satisfy the rules

$$\int A \, f(x) \, dx = A \int f(x) \, dx$$

$$\int \left[f(x) + g(x) \right] dx = \int f(x) \, dx + \int g(x) \, dx \tag{5.6}$$

where A is a constant, and $f(x)$ and $g(x)$ are arbitrary functions of x. This is useful because it allows us to work out the integral of a sum, or difference, of terms (e.g. a polynomial) by knowing the integrals of its constituent parts.

While eqn (5.6) holds for all integrals, there are several general results that apply specifically to definite ones. The first is that an interchange of the order of the limits reverses the sign of the integral, and follows from eqn (5.5)

$$\int_b^a f(x) \, dx = - \int_a^b f(x) \, dx \tag{5.7}$$

Another formula emanating from eqn (5.5) concerns the decomposition of an integral into the sum of two, or more, sharing suitable common limits

$$\int_a^c f(x) \, dx = \int_a^b f(x) \, dx + \int_b^c f(x) \, dx \tag{5.8}$$

This can be understood directly, at least for the case $a \leq b \leq c$, by thinking about the relevant areas under the curve $y = f(x)$. Finally, the formal inverse relationship between integration and differentiation can be exploited to yield

$$\frac{d}{dt} \int_{a(t)}^{b(t)} f(x) \, dx = f(b) \frac{db}{dt} - f(a) \frac{da}{dt} \tag{5.9}$$

If $G(t) = \int_t^{t^2} \sin x \, dx$

$$= \cos t - \cos t^2$$

$$\frac{dG}{dt} = 2t \sin t^2 - \sin t$$

where we have implicitly used eqns (5.4) and (5.5), and the chain rule of eqn (4.12). Usually a and b are constants, and so the definite integral is simply a number (not a function of x, which is said to be a 'dummy variable') whose derivative is equal to zero. If the limits do depend on the value of a parameter, say t, then we would generally expect the area under the curve to vary with respect to it; although the related rate-of-change could be calculated by first evaluating the integral and then differentiating the result with respect to t, eqn (5.9) provides a short cut that avoids the need for explicitly carrying out the initial step and simplifies the latter.

5.4 Inspection and substitution

As noted earlier, most integrals are evaluated by remembering that integration is the reverse of differentiation. Thus the backwards application of the results in chapter 4 gives rise to many 'standard integrals'. For example, eqns (4.7) and (4.8) yield

$$\int e^x \, dx = e^x + C \quad \text{and} \quad \int \frac{1}{x} \, dx = \ln x + C \qquad (5.10)$$

where C is the constant of integration, and x is assumed to be positive in the logarithmic case. The key step in the calculation of $\int f(x) \, dx$ is the recognition of the *integrand*, $f(x)$, as the differential coefficient of a known function; this requires a spotting ability which is enhanced through experience, but initiated by acquiring a good familiarity with the derivatives of the previous chapter. Whenever possible, an integral should be checked by making sure that its derivative returns the integrand; this test eliminates many elementary errors.

$$\int x^M \, dx = \frac{x^{M+1}}{M+1} + C$$

$$(\text{for } M \neq -1)$$

When $\int f(x) \, dx$ does not appear to be of a standard form, it can often be turned into one through a suitable 'change of variables'; that is to say, the *substitution* of $u = g(x)$ may lead to an integral of a function of u which is recognised more easiiy. For example, the integral of $x\sqrt{4-x}$, with respect to x, becomes straightforward on putting $u = 4 - x$

$$\int x\sqrt{4-x} \, dx = -\int (4-u)\sqrt{u} \, du = \int (u^{3/2} - 4u^{1/2}) \, du$$

where the last step makes use of eqns (1.3) and (1.4). While the substitution of $x = 4 - u$ is obvious, the replacement of dx requires a bit of thought; it follows from the derivative $dx/du = -1$, so that $dx = -du$. If this splitting up of dx and du seems strange, then the result can be viewed as the limiting form of the relationship $\delta x \approx dx/du \times \delta u \approx -\delta u$. The linear property of eqn (5.6), and a knowledge of the integral of u^M, then gives

$$\int x\sqrt{4-x} \, dx = \frac{2}{5} u^{5/2} - \frac{8}{3} u^{3/2} + C$$

where the right-hand side can be expressed in terms of x by putting $u = 4 - x$. This final 'back-substitution' is not needed for a definite integral if the limits $x = a$ and $x = b$ are applied in terms of their equivalent values in u; that is, $u = 4 - a$ and $u = 4 - b$.

There are no general rules for working out which substitution to use, or deciding whether one will be helpful at all, without actually trying it. There are several cases that are met frequently, however, and it is worth mentioning these explicitly. Firstly, consider the integral of $\sin^M x \cos^N x$ where M and N are integers. If M is odd, then putting $u = \cos x$ transforms the trigonometric integrand in x to a simple polynomial one in u; remembering, of course, that $du = -\sin x \, dx$ and even powers of $\sin x$ can be factored in terms of $1 - u^2$ from eqn (3.8). Similarly, putting $u = \sin x$ is helpful if N is odd. If both M and N are even, then the integration usually entails a lot more effort. In particular, it involves the repeated use of the double-angle formulae of section 3.4, especially eqn (3.15), to transform the integrand into a familiar form; as an example, try integrating $\sin^4 x$ with the aid of exercise 3.8.

$$\int_0^{\pi/2} \sin^4 x \, \cos^3 x \, dx$$

$$= \int_0^1 u^4 (1 - u^2) \, du$$

where $u = \sin x$

If $t = \tan\left(\frac{x}{2}\right)$ $dx = \frac{2\,dt}{1+t^2}$

$\sin x = \frac{2t}{1+t^2}$ $\cos x = \frac{1-t^2}{1+t^2}$

In integrals containing the term $(a^2 - x^2)$, where a is a constant, it is often helpful to put $x = a\sin\theta$ or $x = a\cos\theta$; with factors of $(a^2 + x^2)$ and $(x^2 - a^2)$, the substitution $x = a\tan\theta$ and $x = a\sec\theta$ respectively is sometimes useful. As a last resort for integrals of trigonometric functions of x, it can be worth trying $t = \tan(x/2)$.

Finally, we should mention the special case where the integrand takes the form of some function of $u = g(x)$ times du/dx. The presence of the derivative simplifies the integral greatly since, according to the chain rule of section 4.5, we then have

$$\int f(u)\,\frac{du}{dx}\,dx = \int f(u)\,du \tag{5.11}$$

$\int \frac{2x}{x^2-1}\,dx = \ln(x^2-1)+C$

If this situation can be spotted then the explicit substitution $u = g(x)$ is not necessary and, with experience, the integral can often be evaluated directly in our head. For example, the integral of $x\exp(-x^2)$ is equal to minus a half of $\exp(-x^2)$ because, to within a factor of $-1/2$, the integrand is the product of 'e to the power of something and the derivative of the something'. We should also note that the prefactor x is crucial in this instance as the integral of $\exp(-x^2)$ itself is not at all straightforward!

5.5   Partial fractions

In section 1.7, we discussed the topic of partial fractions whereby an algebraic ratio is decomposed into the sum of several constituents parts. While this procedure seemed like a backwards step, it can be very useful for evaluating suitable types of integrals. As a simple illustration, consider the case

$$\int \frac{dx}{x(x+1)} = \int \frac{dx}{x} - \int \frac{dx}{x+1}$$

where we have used a partial fraction analysis of $1/[x(x+1)]$, and the linear property of eqn (5.6), to obtain the difference of the two terms on the right-hand side; the latter integrals are easily recognised as being given by the logarithms of their respective denominators.

5.6   Integration by parts

The integration of eqn (4.9), which gives the rule for differentiating a product, leads to the general relationship

$$\int u\,\frac{dv}{dx}\,dx = uv - \int v\,\frac{du}{dx}\,dx \tag{5.12}$$

where u and v are arbitrary functions of x. This slightly odd-looking equation refers to the situation where the integrand consists of the product of two terms of which one, labelled as u, can be differentiated and the other, denoted by dv/dx, easily integrated. The use of eqn (5.12) is called *integration by parts*, with the idea being that the integral on the right-hand side might be more 'doable' than the original one on the left.

As an example, suppose that we wanted to evaluate $\int x\ln x\,dx$. Then, since the integral of x is $x^2/2$ and the derivative of $\ln x$ is $1/x$, eqn (5.12) gives

$$\int x \ln x \, dx = \frac{x^2}{2} \ln x - \int \frac{x}{2} \, dx$$

where the integral on the far right is just $x^2/4 + C$. On a related note, and somewhat surprisingly, the integral of $\ln x$ can be ascertained in this manner by expressing the integrand as $1 \times \ln x$.

$$\int \ln x \, dx = x \ln x - x + C$$

5.7 Reduction formulae

Consider the following integral I_n, where the subscript n corresponds to the power of x in the integrand

$$I_n = \int_0^\infty x^n e^{-x} \, dx \qquad (5.13)$$

If $n = 0$, so that $x^n = 1$, then it is simple to show that $I_0 = 1$. If n is a large positive integer, then I_n can be evaluated by relating it to I_0 through the repeated use of integration by parts. To see this, let's carry out this procedure on I_n just once

$$\int_0^\infty e^{-x} \, dx = \left[-e^{-x}\right]_0^\infty = 1$$

$$I_n = \left[-x^n e^{-x}\right]_0^\infty + n \int_0^\infty x^{n-1} e^{-x} \, dx$$

where we have integrated e^{-x} and differentiated x^n. The first term on the right is equal to zero, because $x^n e^{-x}$ is nought at both the upper and lower limit; by the definition of eqn (5.13), the integral is I_{n-1}. Hence, we have

$$I_n = n I_{n-1} \qquad (5.14)$$

If $n = 7$, for example, then $I_7 = 7I_6$. Using eqn (5.14) several times over, we obtain $I_6 = 6I_5$, $I_5 = 5I_4$, and so on until we reach $I_1 = I_0 = 1$. Hence, on combining all these elements together, we see that I_n is equal to n-factorial. This integral definition of $n!$ is called a *Gamma function*, $\Gamma(n+1)$, and confirms our assertion in section 1.5 that $0! = 1$.

$$\int_0^\infty x^n e^{-x} \, dx = n!$$

The main purpose of the calculation above was to illustrate the derivation and use of a *reduction formula*, such as eqn (5.14); these relate integrals of 'order' n, generally denoted by I_n, to corresponding ones of lower 'powers'.

5.8 Symmetry, tables and numerical integration

In concluding this chapter, we should reiterate that the successful evaluation of an integral hinges on our ability to spot an integrand as the differential coefficient of a known function. Procedures like substitution and integration by parts are often helpful in that they can transform an unfamiliar case into one that is easily recognised; several of these manipulations may be needed in any given problem, and sometimes more than once.

In real life, unlike exams, the best way of proceeding is usually to look up the result in a 'handbook of integrals'. Even with the aid of these reference works, there are many situations when an integral cannot be done; that is to say, it doesn't have a simple algebraic form. A definite integral can always be computed numerically, of course, by calculating the area under the curve by using the summation of eqn (5.1); commonly-met cases, such as the integral of $\exp(-x^2)$ between zero and a variable upper limit (called an *error function*), are frequently found tabulated in books.

Finally, we should mention that some integrals can be set to zero based solely on the symmetry of their integrands. This is especially true of *anti-symmetric* or *odd* functions, like $f(x) = x$ and $f(x) = \sin x$, when they are integrated over a symmetric range around the origin

$$\text{If } f(-x) = -f(x), \quad \int_{-a}^{a} f(x)\,dx = 0 \tag{5.15}$$

This property can either be proved algebraically, with eqns (5.7) and (5.8), or reasoned graphically: as the curve $y = f(x)$ on one side of the line $x = 0$ is an upside-down mirror image of that on the other, the integral is zero because the areas in the two halves have the same magnitude but opposite signs. The counterpart of eqn (5.15) for *symmetric* or *even* functions, like $f(x) = x^2$ and $f(x) = \cos x$, where $f(-x) = f(x)$, is that the integral from $-a$ to a is equal to twice that from 0 to a. Since most functions are neither odd nor even, but are of a mixed symmetry, the above discussion does not apply to them.

Exercises

5.1 Integrate the following functions with respect to x: (i) $x + \sqrt{x} - 1/x$; (ii) $\sqrt{x}(x - 1/x)$; (iii) 2^x; (iv) e^{2x}; (v) $1/(2x - 1)$; (vi) $\sin 2x + \cos 3x$; (vii) $\tan x$; (viii) $\sin^2 x$; (ix) $x/(1+x^2)$; (x) $1/(1+x^2)$.

5.2 Evaluate the following definite integrals:

(i) $\int_0^{\pi/2} \sin^4 x \cos x\,dx$, (ii) $\int_0^{\pi/2} \sin^4 x\,dx$, (iii) $\int_0^4 \frac{x+3}{\sqrt{2x+1}}\,dx$

5.3 Integrate the three partial fractions expressions in exercise 1.12, with respect to x.

5.4 By integrating by parts (twice), show that

(i) $\int x \sin x\,dx = C - x\cos x + \sin x$

(ii) $\int \sin x\, e^{-x}\,dx = C - (\sin x + \cos x)e^{-x}/2$

5.5 If

$$I_n = \int_0^{\pi/2} \sin^n x\,dx,$$

for $n \geq 1$, show that $nI_n = (n-1)I_{n-2}$. Hence, by relating odd orders of I_n to I_1 and even ones to I_0, evaluate I_5 and I_8.

5.6 Prove eqn (5.15) algebraically, and derive the corresponding expression for the symmetric function.

6 Taylor series

6.1 Approximating functions

When dealing with a complicated function, it can be useful to approximate it with one of a simpler form. While the latter may not represent a complete and accurate description of the situation at hand, it frequently provides the only means of making analytical progress. There are many approximations that could be used, of course, but it is the one that captures all the salient features that is most helpful. In this chapter we will focus on the *Taylor series*, which is appropriate when our principle interest lies in the behaviour of a function in the neighbourhood of a particular point.

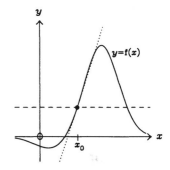

Consider the curve $y = f(x)$. The crudest approximation to this function is a horizontal line $y = a_0$, where a_0 is a constant; if $a_0 = f(x_0)$, then it will even be correct at $x = x_0$. A better approximation would be a sloping line $y = a_0 + a_1(x - x_0)$, where the coefficient a_1 allows for a non-zero gradient; if $a_1 = 0$, then we return to the earlier case. Continuing along this path, we could add a quadratic (or curvature) term $a_2(x - x_0)^2$, a cubic contribution $a_3(x - x_0)^3$, and so on, to gain further improvements. Thus, a function $f(x)$ can be approximated about the point x_0 by using a polynomial expansion

$$f(x) \approx a_0 + a_1(x - x_0) + a_2(x - x_0)^2 + a_3(x - x_0)^3 + \cdots \qquad (6.1)$$

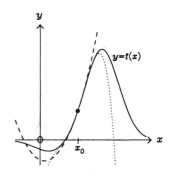

This is, in fact, the essence of a Taylor series. Before dealing with the details of the coefficients, we should point out that the advantage of eqn (6.1) is that its right-hand side is usually easier to calculate, differentiate, integrate, and generally manipulate, than the expression on the left.

6.2 Derivation of the Taylor series

As we shall see shortly, the approximation of eqn (6.1) becomes an equality when enough terms are added and the magnitude of the difference $x - x_0$ is not too large. The substitution $x = x_0$ then gives $f(x_0) = a_0$, because everything involving $x - x_0$ on the right-hand side is equal to zero. If we differentiate eqn (6.1) with respect to x and then put $x = x_0$, we obtain $f'(x_0) = a_1$. Repeating this procedure many times over, we find that $f''(x_0) = 2a_2$, $f'''(x_0) = 6a_3$, and, in general, that the value of the nth derivative of $f(x)$ at $x = x_0$, $f^n(x_0)$, is equal to $n!$ times a_n. Hence, the Taylor series can be written concisely as

$$f(x) = \sum_{n=0}^{\infty} \frac{f^n(x_0)}{n!} (x - x_0)^n \qquad (6.2)$$

An alternative way of expressing this sum, which is sometimes more useful, is obtained by making the substitution $\Delta = x - x_0$; explicitly, this gives

$$f(x_0 + \Delta) = f(x_0) + \Delta f'(x_0) + \frac{\Delta^2}{2!} f''(x_0) + \frac{\Delta^3}{3!} f'''(x_0) + \cdots \qquad (6.3)$$

The special case of eqn (6.3) when $x_0 = 0$ is often called a *Maclaurin series*. Apart from the origin, other common points of expansion for a Taylor series tend to be the locations of the maxima and minima of $f(x)$.

6.3　Some common examples

As a concrete example of the discussion above, let's work out the Maclaurin series for $\sin x$. Such problems are best done by setting up a two-column list with the function and its derivatives on the left, and their corresponding values at $x = x_0$ (in this case zero) on the right

$f(x) =$	$\sin x$	$f(0) =$	0
$f'(x) =$	$\cos x$	$f'(0) =$	1
$f''(x) =$	$-\sin x$	$f''(0) =$	0
$f'''(x) =$	$-\cos x$	$f'''(0) =$	-1

Higher-order derivatives can be ascertained from this table by noting that the fourth one is $\sin x$, and then the whole cycle repeats itself. Substituting for the various $f^n(0)$ in eqn (6.2), we obtain

$$\sin x = x - \frac{x^3}{3!} + \frac{x^5}{5!} - \frac{x^7}{7!} + \cdots \qquad (6.4)$$

$$\sin\left(\frac{1}{2}\right) = \frac{1}{2} - \frac{1}{48} + \frac{1}{3840} - \cdots$$

This provides us with the means of calculating the sine of an angle without having to draw any triangles; to improve the accuracy of the answer, we simply add more terms. The small angle approximation of eqn (3.7) follows immediately when $|x| \ll 1$, where x is measured in radians, because the higher powers of x rapidly become negligible.

A retracing of the steps of the analysis above for $f(x) = \cos x$ yields

$$\cos x = 1 - \frac{x^2}{2!} + \frac{x^4}{4!} - \frac{x^6}{6!} + \cdots \qquad (6.5)$$

This expansion would also have resulted if we had differentiated, or integrated, eqn (6.4) term-by-term. Finally, it is readily shown that the Taylor series of $f(x) = e^x$ is

$$e^x = 1 + x + \frac{x^2}{2!} + \frac{x^3}{3!} + \frac{x^4}{4!} + \frac{x^5}{5!} + \cdots \qquad (6.6)$$

$$e = 1 + 1 + \frac{1}{2} + \frac{1}{6} + \frac{1}{24} + \cdots$$

Again, a term-by-term differentiation of eqn (6.6) indicates how the derivative of an exponential can be equal to the function itself.

6.4　The radius of convergence

The power of a Taylor series lies in the fact that it enables us to approximate an arbitrary function by a simple low-order polynomial in the vicinity of a particular point; for greater accuracy, we just need to extend the expansion to a higher order. The last statement must be qualified in that its validity hinges on the property that the terms omitted from the summation become ever smaller and more negligible. This is only true as long as $|x - x_0| < R$, where the threshold R is known as the *radius of convergence*; the Taylor series should not be used outside this domain. Formally, this convergence criterion requires that the ratio of neighbouring terms in the expansion, higher to lower, is less than unity at the infinite-order end of the series. While $R \to \infty$ for eqns (6.4)–(6.6), so that those three cases are valid for all values of x, the Taylor series for $f(x) = (1+x)^n$ has a very limited range of applicability

$$(1+x)^n = 1 + nx + \frac{n(n-1)}{2!}x^2 + \frac{n(n-1)(n-2)}{3!}x^3 + \cdots \quad (6.7)$$

for $|x| < 1$. This is a generalised version of the binomial expansion met in section 1.5, where the power need not be a positive integer; if n is a positive integer then the right-hand side of eqn (6.7) terminates (or becomes zero) after a finite number of terms, so that eqn (1.11) is recovered, and the expansion is valid for all values for x. Another Taylor series that is encountered frequently is

$$\sqrt{1+x} = 1 + \frac{x}{2} - \frac{x^2}{8} + \cdots$$

$$\ln(1+x) = x - \frac{x^2}{2} + \frac{x^3}{3} - \frac{x^4}{4} + \cdots \quad (6.8)$$

which converges as long as $|x| < 1$.

6.5 L'Hospital's rule

The Taylor series is a valuable tool for ascertaining limits in awkward situations. For example, the value of $\sin x / x$ is not obvious when $x = 0$; in essence, it's equivalent to trying to work out the product of nought and infinity! The desired ratio can be obtained by considering how the numerator and denominator behave as they gradually approach zero; in other words, by using a Taylor series expansion about the origin

$$\lim_{x \to 0} \frac{\sin x}{x} = \lim_{x \to 0} \frac{x - x^3/6 + \cdots}{x} = \lim_{x \to 0} 1 - \frac{x^2}{6} + \cdots$$

the required result is seen to be unity.

Extending the arguments of the specific case above to the general ratio $h(x)/g(x)$, by expanding the numerator and the denominator in Taylor series about $x = a$, it is easily shown that

$$\lim_{x \to a} \frac{h(x)}{g(x)} = \lim_{x \to a} \frac{h'(x)}{g'(x)} \quad (6.9)$$

if both $h(a)$ and $g(a)$ are zero. If the first derivatives, $h'(a)$ and $g'(a)$, are also nought, then the desired result is given by the ratio of the second derivatives, $h''(a)/g''(a)$; and so on until a well-defined limit is obtained. This is known as *l'Hospital's rule*.

$$\lim_{x \to 0} \frac{\sin x}{x} = \lim_{x \to 0} \frac{\cos x}{1} = 1$$

6.6 Newton–Raphson algorithm

The Taylor series also provide a powerful means of finding the roots of an equation numerically; that is to say, ascertaining the values of x which satisfy the relationship $f(x) = 0$. For an arbitrary function, $f(x)$, the solutions may be difficult to obtain algebraically. Nevertheless we may know that $x = x_0$ is a good guess, so that $f(x_0) \approx 0$; how can we use this to derive a better estimate? Well, expanding $f(x)$ in a Taylor series about $x = x_0$ we have

$$f(x) = f(x_0) + (x - x_0)f'(x_0) + \cdots = 0$$

Assuming that our initial guess is good enough that the quadratic and higher-order terms are negligible by comparison with the zeroth and first-order ones, the resulting linear equations can readily be solved to yield

$$x \approx x_0 - \frac{f(x_0)}{f'(x_0)} \tag{6.10}$$

In other words, the values of the function and its first derivative at x_0 enable us to obtain a better estimate of the root according to eqn (6.10). This can, of course, be used as our new initial guess, and the procedure repeated until there is no significant change between x and x_0; this *iterative* numerical method of solving $f(x) = 0$ is known as the *Newton-Raphson* algorithm.

Exercises

6.1 Derive the Taylor series for $\cos x$, e^x, and $\sin(x + \pi/6)$ for small x.

6.2 Derive the binomial expansion of eqn (6.7). By writing $\sqrt{C}$ as $a(1 + b)^{1/2}$, where a and b are suitable constants, find $\sqrt{8}$ and $\sqrt{17}$ to four decimal places.

6.3 Derive the Taylor series for $\ln(1 + x)$ and $\ln(1 - x)$. Hence, write down the power series for $\ln[(1 + x)/(1 - x)]$.

6.4 Determine the following limits

$$(a) \quad \lim_{x \to 0} \frac{\sin(ax)}{x}, \qquad (b) \quad \lim_{x \to 0} \frac{\cos x - 1}{x},$$

$$(c) \quad \lim_{x \to 0} \frac{2\cos x + x \sin x - 2}{x^4}.$$

6.5 Given that $x^3 + 3x^2 + 6x - 3 = 0$ has only one (real) root, find its value to three significant figures by using the Newton-Raphson method.

7 Complex numbers

7.1 Definition

If any number, integer or fraction, positive or negative, is multiplied by itself, then the result is always greater than, or equal, to zero. What, then, is the square root of -9? To address this question we need to invent an *imaginary* number, usually denoted by 'i', whose square is defined to be negative

$$i^2 = -1 \qquad (7.1)$$

A *real* number, say b (where $b^2 \geq 0$), times i is also imaginary; it's just b times bigger than i. If a is also an ordinary number, then the sum z of a and ib

$$z = a + ib \qquad (7.2)$$

is known as a *complex* number; this does not indicate an intrinsic difficulty with the concept, but highlights the hybrid nature of the entity. It consists of both a real part and an imaginary one

$$\mathcal{R}e\{z\} = a \quad \text{and} \quad Im\{z\} = b \qquad (7.3)$$

It may seem odd that $Im\{z\}$ is b rather than ib, but this is so because it represents the size of the imaginary component.

    Although the construct of imaginary numbers appears arbitrary, complex numbers turn out to be very useful for tackling many real-life problems. Our aims here, however, are to acquire a knowledge of their basic properties.

7.2 Basic algebra

Let's start with the most elementary operations, namely addition and subtraction. To add, or subtract, complex numbers, we simply combine the real and imaginary parts separately. For example

$$a + ib \pm (c + id) = a \pm c + i(b \pm d) \qquad (7.4)$$

where a, b, c, and d are real. In fact we have made an implicit assumption in writing the term on the far right, in that complex numbers obey the same rules as ordinary ones; but with every occurrence of i^2 replaced by -1. Thus, it is easy to show that the product of $a + ib$ and $c + id$ is given by

$$1 + 2i - (5 - i) = -4 + 3i$$

$$(a + ib)(c + id) = ac - bd + i(ad + bc) \qquad (7.5)$$

because $i^2 bd = -bd$. Division is less straightforward in that it entails the use of the *complex conjugate* of the denominator; let us consider this first.

$$(1 + 2i)(3 - i) = 5 + 5i$$

    The conjugate of a complex number z, denoted by z^*, is defined to have the same real part but the opposite imaginary component; that is, $Re\{z^*\} = Re\{z\}$ and $Im\{z^*\} = -Im\{z\}$. In terms of eqn (7.2), therefore

$$z^* = (a+ib)^* = a-ib \qquad (7.6)$$

Thus, complex numbers and their conjugates satisfy the following relationships

$$\begin{aligned} z+z^* &= 2a &= 2\,Re\{z\} \\ z-z^* &= 2ib &= 2i\,Im\{z\} \\ zz^* &= a^2+b^2 &= |z|^2 \end{aligned} \qquad (7.7)$$

We will come to the meaning of $|z|$ shortly, but the important point about eqn (7.7) is that the product $zz^* = a^2 + b^2$ is a real number because it does not involve any i's. This feature enables us to calculate the real and imaginary part of the ratio of two complex numbers by multiplying both the top and bottom of the quotient by the conjugate of the denominator

$$\frac{a+ib}{c+id} = \frac{a+ib}{c+id} \times \frac{c-id}{c-id} = \frac{ac+bd+i(bc-ad)}{c^2+d^2} \qquad (7.8)$$

To evaluate the ratio $(1+2i)/(3-i)$, for example, we multiply it by unity in the form $(3+i)/(3+i)$; this gives a real denominator of $3^2 + 1^2 = 10$, and a complex numerator of $1 + 7i$. Hence the result is $1/10 + i7/10$.

7.3 The Argand diagram

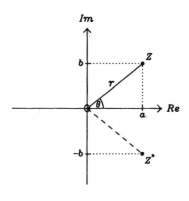

So far we have considered complex numbers from an algebraic point-of-view; it is often helpful to think of them in geometrical terms. This is easily done with the aid of an *Argand diagram* where the horizontal, or x, axis of a graph is seen as representing the real part of a complex number, and the vertical, or y, axis gives the imaginary component. Thus the point with (x,y) coordinates (a,b) corresponds to the complex number $z = a+ib$. Its conjugate z^* is a reflection in the real axis. An alternative way of specifying the location of a point on a graph is through its distance r from the origin, and the anticlockwise angle θ that this 'radius' makes with the (positive) real axis. In this system r is known as the *modulus*, *magnitude* or *amplitude* of z; θ is called its *argument* or *phase*.

By using the elementary trigonometry met in section 3.2, we can relate the quantities r, θ, a and b in the Argand diagram by

$$a = r\cos\theta \qquad \text{and} \qquad b = r\sin\theta \qquad (7.9)$$

or, in the reverse sense, through

$$r^2 = a^2 + b^2 \qquad \text{and} \qquad \theta = \tan^{-1}(b/a) \qquad (7.10)$$

A comparison between eqns (7.7) and (7.10) shows that $zz^* = r^2$, where $r = |z|$ is the modulus of the complex number. We need to qualify the second part of eqn (7.10) in that there is ambiguity in $\tan^{-1}(b/a)$. If $z = -1-i$, for example, then a plot of this point in the Argand diagram shows θ to be $5\pi/4$ radians or $225°$; although this is a value of $\tan^{-1}(1)$, so too is $45°$ which does not give the correct argument. It is also worth remembering that θ is only defined to within a factor of 2π, because we could add (or subtract) any integer number of $360°$ to it and obtain the same point in the Argand diagram.

7.4 The imaginary exponential

Perhaps the single most important result in complex analysis concerns the exponential of an imaginary number

$$e^{i\theta} = \cos\theta + i\sin\theta \qquad (7.11)$$

While this equation is not immediately obvious, it is easily verified by substituting $x = i\theta$ in the Taylor series expansion of eqn (6.6) and collecting together the odd and even powers of θ separately; a comparison with eqns (6.4) and (6.5), while remembering that $i^2 = -1$, then yields eqn (7.11). The product of r and $e^{i\theta}$ allows a complex number to be expressed in a very compact form in terms of its modulus and argument:

$$a + ib = r(\cos\theta + i\sin\theta) = re^{i\theta} \qquad (7.12)$$

where a, b, r, and θ are related through eqns (7.9) and (7.10).

$$
\begin{aligned}
e^{i\pi/4} &= (1+i)/\sqrt{2} \\
e^{i\pi/2} &= i \\
e^{i3\pi/4} &= (-1+i)/\sqrt{2} \\
e^{i\pi} &= -1
\end{aligned}
$$

As we shall see shortly, the exponential form of a complex number is very useful when dealing with roots and logarithms; it also provides a valuable insight into products and quotients. Consider two complex numbers z_1 and z_2 having amplitudes r_1 and r_2, and phases θ_1 and θ_2

$$z_1 = r_1 e^{i\theta_1} \quad \text{and} \quad z_2 = r_2 e^{i\theta_2}$$

Then, using the rules for combining indices in section 1.2, we obtain

$$z_1 z_2 = r_1 r_2 e^{i(\theta_1 + \theta_2)} \qquad (7.13)$$

In other words, by comparing the right-hand side with the standard form $re^{i\theta}$, we see that the modulus of a product is equal to the product of the moduli; and that the argument of a product is the sum of the arguments. Similarly, the modulus of a quotient is equal to the ratio of the moduli, and its argument is the difference of the arguments

$$\frac{z_1}{z_2} = \frac{r_1}{r_2} e^{i(\theta_1 - \theta_2)} \qquad (7.14)$$

The product of a complex number with its conjugate is a special case of eqn (7.13) with $z_2 = z_1^*$, so that $r_2 = r_1$ and $\theta_2 = -\theta_1$, and immediately returns the now familiar result that $zz^* = |z|^2$.

7.5 Roots and logarithms

At the beginning of this chapter we posed the question of the square root of -9. While the answer to this simple problem may now be intuitively obvious, as $\pm 3i$, how can we work it out in a systematic fashion? Well, technically, we wish to find the value of the complex number z which satisfies the equation $z^2 = -9$. This task becomes easier if we write the right-hand side in the form $re^{i\theta}$. For -9, the amplitude is 9 and the phase is $180°$, give or take any integer multiple of $360°$; thus we have

$$z^2 = -9 = 9\,e^{i(\pi \pm 2n\pi)}$$

where n is any integer (and θ is in radians). Raising both sides to the power of a half, according to the rules in section 1.2, gives the square root as

$$z = 9^{1/2} e^{i(\pi \pm 2n\pi)/2} = 3 e^{i(\pi/2 \pm n\pi)}$$

Hence, an even value of n (such as $n = 0$) yields the solution $z = 3i$, whereas an odd n (like $n = 1$) gives the alternative result of $z = -3i$.

As a second example of this type of calculation, let's consider the cube roots of i; that is, the solution of the equation $z^3 = i$. Again, the best procedure is to write the right-hand side in the form $re^{i\theta}$

$$z^3 = i = e^{i(\pi/2 \pm 2n\pi)}$$

where n is any integer. Raising both sides to the power of a third, we obtain

$$z = e^{i(\pi/2 \pm 2n\pi)/3} = e^{i(\pi/6 \pm 2n\pi/3)}$$

Three distinct solutions emerge for $n = 0$, $n = 1$ and $n = 2$; namely $(\sqrt{3} + i)/2$, $(-\sqrt{3} + i)/2$ and $-i$ respectively. All other values of n keep generating the same results (e.g. $n = 3$ is identical to $n = 0$).

If the cube roots of i are plotted on an Argand diagram, they are seen to lie at the vertices of an equilateral triangle. Such a uniform distribution of solutions around the circumference of a circle is a general feature of this type of problem, with the m-roots of a complex number leading to an m-sided regular polygon.

In addition to its use for roots and powers, the $re^{i\theta}$ form of a complex number allows us to extend the domain of logarithms. When restricted to real numbers, a logarithm could only be defined for positive quantities; now, using the rules of section 1.2, we can take the logarithm of anything

$$\ln(re^{i\theta}) = \ln(r) + i\theta \qquad (7.15)$$

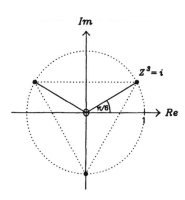

$$\ln(2i) = \ln\left[2 e^{i(\pi/2 \pm n\pi)}\right]$$

$$= \ln 2 + i(\pi/2 \pm n\pi)$$

The amplitude r is positive by definition, and the phase θ is usually taken as being zero for real numbers. Since we know that θ entails an ambiguity with respect to the addition of any integer numbers of 2π radians, the imaginary part of a logarithm can take an infinite, but discrete, number of values; by convention, the *principal* one is chosen to lie between $-\pi$ and $+\pi$ radians.

7.6   De Moivre's theorem and trigonometry

A useful trigonometric result also follows from the imaginary exponential of section 7.4 when it is combined with eqn (1.5)

$$\left(e^{i\theta}\right)^m = e^{im\theta}$$

Using the definition of eqn (7.11) on both sides separately, we obtain *de Moivre* theorem

$$(\cos\theta + i\sin\theta)^m = \cos(m\theta) + i\sin(m\theta) \qquad (7.16)$$

This yields, for example, the formulae for the sines and cosines of multiple angles. As a simple illustration, consider the case of $m = 2$

$$\cos(2\theta) + i\sin(2\theta) = \cos^2\theta + 2i\sin\theta\cos\theta - \sin^2\theta$$

which, on equating the real and imaginary parts, returns the double-angle formulae of eqns (3.14) and (3.11) respectively.

Further connections between the imaginary exponential and trigonometry emerge if we combine eqn (7.11) with its complex conjugate

$$e^{-i\theta} = \cos\theta - i\sin\theta$$

Their sum gives a formula for $\cos\theta$, while the difference yields $\sin\theta$

$$\sin\theta = \frac{e^{i\theta} - e^{-i\theta}}{2i} \quad \text{and} \quad \cos\theta = \frac{e^{i\theta} + e^{-i\theta}}{2} \qquad (7.17)$$

The tangent is obviously given by the ratio of these expressions, as defined in eqn (3.3). Amongst other things, these relationships are useful for rewriting powers of sines and cosines in terms of their mutiple-angle equivalents. To take a very simple example

$$\sin^2\theta = \left(\frac{e^{i\theta} - e^{-i\theta}}{2i}\right)^2 = \frac{e^{2i\theta} - 2 + e^{-2i\theta}}{-4} = \frac{1}{2}\left[1 - \left(\frac{e^{2i\theta} + e^{-2i\theta}}{2}\right)\right]$$

which can be recognised as being $(1 - \cos 2\theta)/2$, as in eqn (3.15).

7.7 Hyperbolic functions

All the trigonometric functions we met in chapter 3 have a set of counterparts known as *hyperbolic* (trigonometric) functions; they are designated by having an 'h' appended to the abbreviations of the former. For example, sinh, cosh and tanh are the hyperbolic sin, cos and tan respectively. We will see the relevance to complex numbers shortly, but begin with the definition of $\sinh x$ and $\cosh x$

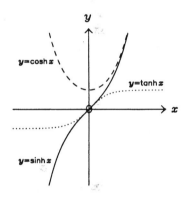

$$\sinh x = \frac{e^x - e^{-x}}{2} \quad \text{and} \quad \cosh x = \frac{e^x + e^{-x}}{2} \qquad (7.18)$$

By analogy with eqn (3.3), the hyperbolic tangent is the ratio of sinh and cosh

$$\tanh x = \frac{\sinh x}{\cosh x} = \frac{e^x - e^{-x}}{e^x + e^{-x}} = \frac{e^{2x} - 1}{e^{2x} + 1} \qquad (7.19)$$

The shapes of the curves $y = \sinh x$, $\cosh x$ and $\tanh x$ can be ascertained by considering the characteristics of the exponential function met in section 2.4; in particular, by looking at the behaviour of $y = e^x$ and $y = e^{-x}$ in the limits of $x \to \pm\infty$ and at $x = 0$.

The relationship between hyperbolic and ordinary (circular) trigonometric functions becomes obvious if we substitute $\theta = ix$ into eqn (7.17) and compare the result with eqn (7.18); it yields

$$\sin(ix) = i\sinh(x) \quad \text{and} \quad \cos(ix) = \cosh(x) \qquad (7.20)$$

In other words, the hyperbolic functions are related to the sines, cosines, and so on, of imaginary angles. By using eqn (7.20), we can derive identities for hyperbolic functions which are equivalent to those discussed in section 3.3. For example, since $\sin^2(ix) + \cos^2(ix) = 1$, according to eqn (3.8), we have

$$\cosh^2 x - \sinh^2 x = 1 \qquad (7.21)$$

Indeed, this sort of manipulation leads to an observation, known as *Osborne's rule*, that the identities for hyperbolic functions are the same as those for ordinary trigonometric ones except the sign of every $\sinh^2 x$, or implied $\sinh^2 x$ (as in $\tanh^2 x = \sinh^2 x/\cosh^2 x$) is reversed.

$$1 - \tanh^2 x = \operatorname{sech}^2 x$$

$$1 - \coth^2 x = -\operatorname{cosech}^2 x$$

The derivatives, and integrals, of hyperbolic functions are also readily ascertained by exploiting the convenient differential properties of the exponential function. Thus, using the definition of eqn (7.18), and the material in chapter 4, it is easily shown that

$$\frac{d}{dx}(\sinh x) = \cosh x \quad \text{and} \quad \frac{d}{dx}(\cosh x) = \sinh x \qquad (7.22)$$

$$\frac{d}{dx}(\tanh x) = \operatorname{sech}^2 x$$

and so on. The derivatives of the inverse hyperbolic functions can be obtained in an analogous manner to those of the inverse trigonometric functions. That is to say, by differentiating $x = \sinh y$ with respect to x, and then using eqn (7.20), to yield dy/dx where $y = \sinh^{-1} x$

$$\frac{d}{dx}(\sinh^{-1} x) = \frac{1}{\sqrt{1+x^2}} \qquad (7.23)$$

Incidentally, explicit formulae for the inverse hyperbolic functions, like

$$\frac{d}{dx}(\cosh^{-1} x) = \frac{1}{\sqrt{x^2-1}}$$

$$\frac{d}{dx}(\tanh^{-1} x) = \frac{1}{1-x^2}$$

$$\tanh^{-1} x = \frac{1}{2}\ln\left(\frac{1+x}{1-x}\right) \qquad (7.24)$$

where $|x| \le 1$ (because $|\tanh x| \le 1$) can be derived from the exponential definitions of eqns (7.18) and (7.19). The derivative of $\tanh^{-1} x$ can then be confirmed by differentiating eqn (7.24) with respect to x.

7.8 Some useful properties

We mentioned earlier that complex numbers provide a powerful analytical tool for tackling many advanced theoretical problems. While our examples here are necessarily of an elementary nature, we should specifically state the properties of complex numbers that often make them so useful. Namely, the exponential decomposition of eqn (7.11) and the linearity with respect to addition and subtraction. That is to say that the 'the sum of the real parts is equal to the real part of the sum', or put mathematically

$$\sum_{k=1}^{N} \mathcal{R}e\{z_k\} = \mathcal{R}e\left\{\sum_{k=1}^{N} z_k\right\} \qquad (7.25)$$

where z_k is the kth complex number, and $k = 1, 2, 3, \ldots, N$. The same is true of the imaginary components, of course, and differences. If z is a complex number which is a function of t, say, such as $t^2 + i\cos(t)$, then the linearity of eqn (7.4) leads to the properties

$$\sum_{k=0}^{\infty} \frac{\sin k\theta}{k!} = \sum_{k=0}^{\infty} Im\left\{\frac{e^{ik\theta}}{k!}\right\}$$

$$= Im\left\{\sum_{k=0}^{\infty} \frac{(e^{i\theta})^k}{k!}\right\}$$

$$\frac{d}{dt}(\mathcal{R}e\{z\}) = \mathcal{R}e\left\{\frac{dz}{dt}\right\} \quad \text{and} \quad \int \mathcal{R}e\{z\}\,dt = \mathcal{R}e\left\{\int z\,dt\right\} \qquad (7.26)$$

Again, the same goes for the imaginary parts.

Equations (7.25) and (7.26) are useful because their right-hand sides are frequently easier to evaluate than the corresponding expressions on the left. Let us illustrate this with an example involving integration

$$I = \int e^{at}\cos(bt)\,dt \qquad (7.27)$$

where a and b are real. This problem can be solved by using integration-by-parts twice, as in section 5.6, or by using complex numbers. The latter requires us to spot that the integrand can be written in the form

$$e^{at}\cos(bt) = e^{at}\,\mathcal{R}e\left\{e^{ibt}\right\} = \mathcal{R}e\left\{e^{(a+ib)t}\right\} \tag{7.28}$$

Hence, on using the property of eqn (7.26), we are left with an easy integral

$$I = \mathcal{R}e\left\{\int e^{(a+ib)t}\,dt\right\} = \mathcal{R}e\left\{\frac{e^{(a+ib)t}}{a+ib}\right\} + C \tag{7.29}$$

where C, the constant of integration, is real. While a little effort is required to ascertain the real part of the quotient in eqn (7.29), starting with multiplication of the top and bottom by the complex conjugate of the denominator $(a - ib)$, it is not very difficult to show that

$$\int e^{at}\cos(bt)\,dt = \frac{e^{at}}{a^2+b^2}\left[a\cos(bt) + b\sin(bt)\right] + C$$

or to notice that the imaginary part yields the integral of $e^{at}\sin(bt)$.

Exercises

7.1 If $z = 2+3i$, what is: (a) $\mathcal{R}e\{z\}$, (b) $Im\{z\}$, (c) z^*, (d) $z-z^*$, (e) $z+z^*$, (f) z^2, (g) zz^*?

7.2 If $u = 2+3i$ and $v = 1-i$, find the real and imaginary parts of: (a) $u+v$, (b) $u - v$, (c) uv, (d) u/v, (e) v/u.

7.3 For the previous question, find: (a) $|u|$, (b) $|v|$, (c) $|uv|$, (d) $|u/v|$, (e) $|v/u|$.

7.4 Find the modulus and argument (in the range $-\pi$ to π) of the following, and sketch their positions in an Argand diagram: (a) 1, (b) $1+i$, (c) i, (d) $-1+i$, (e) -1, (f) $1-i$, (g) $-i$, (h) $-1-i$.

7.5 If $z = 1+i\sqrt{3}$, sketch the following in an Argand diagram: (a) z, (b) z^*, (c) z^2, (d) z^3, (e) iz, (f) $1/z$.

7.6 Solve the quadratic equation: $z^2 - z + 1 = 0$.

7.7 Solve the following equations: (a) $z^5 = 1$, (b) $z^5 = 1+i$, (c) $(z+1)^5 = 1$, (d) $(z+1)^5 = z^5$. Sketch the solutions for (a) on an Argand diagram.

7.8 By considering e^{iA} and e^{iB}, derive the compound-angle formulae for $\sin(A+B)$ and $\cos(A+B)$ met in section 3.4.

7.9 Use de Moivre's theorem to find expansions of $\sin 4\theta$ and $\cos 4\theta$ in terms of powers of $\sin\theta$ and $\cos\theta$.

7.10 Show that $\cos^6\theta = (\cos 6\theta + 6\cos 4\theta + 15\cos 2\theta + 10)/32$.

7.11 Use the definitions of eqn (7.18) to show that

$$\sinh(x+y) = \sinh x \cosh y + \cosh x \sinh y$$

and derive a similar expansion for $\cosh(x+y)$.

7.12 Find: (a) $\displaystyle\sum_{k=0}^{\infty}\frac{\cos k\theta}{k!}$, (b) $\displaystyle\int e^{ax}\sin(bx)\,dx$.

8 Vectors

8.1 Definition and Cartesian coordinates

The simplest way to visualise a *vector* is to hold a pencil up in the air. The length of the pencil is a *scalar* quantity, a single number with an associated *unit*, 0.05 metres, say, for a typical pencil. By describing the pencil by a vector, we can give extra information: we can convey not only its length, but also where it is pointing. Vectors are therefore defined by both a *magnitude* (or *modulus*), a positive number, normally with an associated unit, and a *direction* in space. Many physical quantities, like position and velocity, are vectors which must be manipulated in slightly different ways to scalars such as mass, length, and time.

We will use underlining, as in $\underset{\sim}{X}$, rather than bold script to identify vectors and denote their modulus, or length, by $|\underset{\sim}{X}|$; those with a magnitude of 1, such as $\underset{\sim}{X}/|\underset{\sim}{X}|$, are called *unit vectors*.

Unlike scalars, more than one number is needed to describe a vector. Let us start thinking about this by confining our pencil to lie in a plane, for example by dropping it onto a two-dimensional surface like a table. Only two numbers are then required to define its length and orientation, namely the displacements in the x and y directions required to get from one end of the pencil to the other. These offsets are called the *components* of the vector, and are scalar quantities that tell us how far to move along unit vectors parallel with the x and y axes. That is, placing the *origin* of our system at the blunt end of the pencil, we can express the *position vector* of the sharp end by $x\underset{\sim}{i} + y\underset{\sim}{j}$ where we have introduced perpendicular unit vectors $\underset{\sim}{i}$ and $\underset{\sim}{j}$ which point along the x and y axes respectively.

Now lift the sharp end of the pencil off the table, leaving the blunt end at the origin. To describe its length and orientation we now need a third component in the direction of the z-axis, another unit vector $\underset{\sim}{k}$ at right-angles to both $\underset{\sim}{i}$ and $\underset{\sim}{j}$. Mathematically, we say that we have moved from a two-dimensional to a three-dimensional *vector space*. The position vector of the sharp end is now $\underset{\sim}{X} = x\underset{\sim}{i} + y\underset{\sim}{j} + z\underset{\sim}{k}$, which is written in shorthand as (x, y, z). Pythagoras' theorem then allows us to find the modulus-squared of this vector in terms of the components

$$|\underset{\sim}{X}|^2 = x^2 + y^2 + z^2 \tag{8.1}$$

A *suffix notation* can also be used to describe all three components of a vector in the compact form x_m, with m free to take the values 1, 2, and 3, so that $x_1 = x$, $x_2 = y$, and $x_3 = z$.

We call the mutually-perpendicular unit vectors $\underset{\sim}{i}, \underset{\sim}{j}$ and $\underset{\sim}{k}$ the *basis* of the *Cartesian* coordinate system. By convention, the $\underset{\sim}{k}$ vector points upwards at

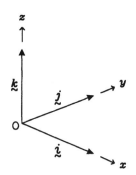

us if we imagine ourselves peering down onto a standard *x–y* plane, or graph paper. It's helpful to train the right-hand to ape these $\underset{\sim}{i}$, $\underset{\sim}{j}$, and $\underset{\sim}{k}$ axes using the thumb, first- and second-fingers respectively.

Although we will confine ourselves to three-dimensional problems, it is worth mentioning that many, but not all, of the ways in which we will learn to manipulate vectors are applicable to *N*-dimensional spaces, where all integer values higher than $N = 3$ are possible. Indeed, as a result of the power of modern computers, scientists and engineers commonly manipulate million-dimensional ($N = 10^6$) vectors. The components are saved in computer memory as a set of *N* scalar values $(a_1, a_2, \ldots, a_N)$.

8.2 Addition, subtraction and multiplication by scalars

As well as having a modulus and a direction, vectors obey some simple rules. The *addition* of vectors $\underset{\sim}{a}$ and $\underset{\sim}{b}$ to form a new vector $\underset{\sim}{c}$ is written as $\underset{\sim}{c} = \underset{\sim}{a} + \underset{\sim}{b}$. Since two vectors are equal only if all of their components are equal, $c_m = a_m + b_m$, for all possible values of *m*; which, in a three-dimensional space, is a shorthand way of expressing three equations.

Subtraction of $\underset{\sim}{b}$ from $\underset{\sim}{a}$ is achieved by reversing the direction of $\underset{\sim}{b}$ and adding: $\underset{\sim}{d} = \underset{\sim}{a} + (-\underset{\sim}{b})$, so that $d_m = a_m - b_m$. In the triangle formed by the origin and the points *A* and *B*, with position vectors $\underset{\sim}{a}$ and $\underset{\sim}{b}$, the vector $\underset{\sim}{d} = \underset{\sim}{a} - \underset{\sim}{b}$ is the one that joins *B* to *A*; this is sometimes written as $\overrightarrow{AB}$.

Multiplication of vectors by scalars is also straightforward: if $\underset{\sim}{b} = \lambda \underset{\sim}{a}$ then the direction of $\underset{\sim}{b}$ is the same as that of $\underset{\sim}{a}$ if λ is positive, or opposite to $\underset{\sim}{a}$ if it is negative; in either case, the modulus of $\underset{\sim}{b}$ is $|\lambda|$-times that of $\underset{\sim}{a}$. An important example concerns the position vector of the mid-point of $\overrightarrow{AB}$, namely

$$\underset{\sim}{e} = \underset{\sim}{a} + \underset{\sim}{d}/2 = \underset{\sim}{a} + (\underset{\sim}{b} - \underset{\sim}{a})/2 = (\underset{\sim}{a} + \underset{\sim}{b})/2$$

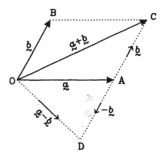

Armed with these basic rules we can now begin to use the general position vector $\underset{\sim}{r}$ to describe the *loci* of points on lines, curves and surfaces within a three-dimensional space. Our first example is

$$\underset{\sim}{r} = \underset{\sim}{a} + \lambda(\underset{\sim}{b} - \underset{\sim}{a}) \tag{8.2}$$

which is the *vector equation* of the straight line which passes through points *A* and *B*: the first term takes us from the origin to a specific point (*A*) on the line, and the second term takes us an arbitrary distance (λ) along the direction fixed by the vector joining *A* to *B*. If $\underset{\sim}{a}$ and $\underset{\sim}{b}$ are the known position vectors of *A* and *B*, then λ can be found in terms of each component, (x, y, z), of $\underset{\sim}{r}$; the *Cartesian equation* of the line is the result of equating these values of λ.

Our second example is

$$\underset{\sim}{r} = \underset{\sim}{a} + \mu(\underset{\sim}{b} - \underset{\sim}{a}) + \nu(\underset{\sim}{c} - \underset{\sim}{a}) \tag{8.3}$$

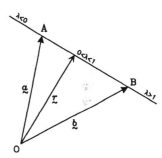

$$\underset{\sim}{r} = (1, 2, 3) + \lambda(3, 2, 1)$$

$$\lambda = \frac{x - 1}{3} = \frac{y - 2}{2} = \frac{z - 3}{1}$$

which is the vector equation of a plane containing the points *A*, *B*, and *C*. The first term takes us from the origin to point *A* on the plane, and the second and third allow us to move arbitrary distances (μ and ν) along the two directions fixed by the vectors joining *A* to *B*, and *A* to *C*. Provided the three points (*A*, *B* and *C*) do not lie along a single line, that is they are not *co-linear*, the vectors $\underset{\sim}{b} - \underset{\sim}{a}$ and $\underset{\sim}{c} - \underset{\sim}{a}$ are said to *span* the two-dimensional space defined by the plane. This means that by choosing suitable values of μ and ν we can get anywhere we like on the plane.

Our final example is

$$|r - a| = R \qquad (8.4)$$

which describes a sphere of radius R centred at the point A: to visualise this note that $(r - a)$ joins A to the general point, and that the modulus sign requires that the length of this vector is always equal to the fixed radius R.

8.3 Scalar product

The *scalar* or *dot product* is one way of 'multiplying vectors together'. It is defined by

$$a \cdot b = |a||b|\cos\theta \qquad (8.5)$$

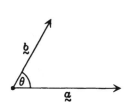

where θ is the angle between the vectors a and b. Since $\cos(-\theta) = \cos(\theta)$ there is no need for any convention to determine 'which way round' to measure the angle, and the result is a simple scalar. If b is taken to be a unit vector, then the geometric meaning of the dot product becomes clear: it is the length $|a|\cos\theta$ which is the *projection* of a in the direction of b. If we take a to be unit instead, then $a \cdot b$ can be viewed as the projection of b along the direction of a. Thus dot products play an important rôle when vectors, such as forces, need to be resolved along particular directions.

If $\theta = \pi/2$ then the dot product is zero, and the vectors a and b are said to be *orthogonal*. The sign of the dot product is positive if θ is acute, and negative if it's obtuse. If $\theta = 0$, the vectors a and b are parallel so that $a \cdot b = |a||b|$. In particular $a \cdot a = |a|^2$, which gives a useful way of finding the modulus (squared) of a vector. Also

$$|b - c|^2 = (b - c) \cdot (b - c) = |b|^2 + |c|^2 - 2b \cdot c \qquad (8.6)$$

because we are allowed to 'multiply the brackets out' just as in normal algebra. In fact, eqn (8.6) provides a remarkably compact proof of the cosine-rule for triangles that we introduced in eqn (3.23).

We next need to find a way of evaluating the dot product in the Cartesian system. It is given by

$$a \cdot b = (a_1 i + a_2 j + a_3 k) \cdot (b_1 i + b_2 j + b_3 k) = a_1 b_1 + a_2 b_2 + a_3 b_3 \qquad (8.7)$$

where we have again multiplied out, and noted that $i \cdot j = j \cdot k = k \cdot i = 0$, and that, being unit vectors, $|i|^2 = |j|^2 = |k|^2 = 1$. Hence the dot product is calculated by summing the products of corresponding components.

Taking the dot product of a vector equation is useful because it produces a scalar equation which can be manipulated according to all the normal algebraic rules. Division by a vector is not defined, however, so never do it!

8.4 Vector product

Another way of multiplying together a and b is with a *vector*, or *cross*, product whose result is also a vector

$$\underset{\sim}{a} \times \underset{\sim}{b} = |\underset{\sim}{a}||\underset{\sim}{b}| \sin\theta \, \underset{\sim}{u} \qquad (8.8)$$

The magnitude of the cross product is equal to $|\underset{\sim}{a}||\underset{\sim}{b}| \sin\theta$, where θ is again the angle between $\underset{\sim}{a}$ and $\underset{\sim}{b}$, and its direction is specified by the unit vector $\underset{\sim}{u}$. The orientation of $\underset{\sim}{u}$ is perpendicular to both $\underset{\sim}{a}$ and $\underset{\sim}{b}$, and hence *normal* to the plane containing them, being uniquely defined by the 'right-hand screw rule'. That is to say, if the right hand is held as though it were gripping a screw-driver, and the curl of the fingers is made to indicate the sense of rotation needed to go from $\underset{\sim}{a}$ to $\underset{\sim}{b}$, then the out-stretched thumb points along $\underset{\sim}{u}$. This prescription immediately tells us that the direction of $\underset{\sim}{a} \times \underset{\sim}{b}$ is opposite to that of $\underset{\sim}{b} \times \underset{\sim}{a}$

$$\underset{\sim}{a} \times \underset{\sim}{b} = -\underset{\sim}{b} \times \underset{\sim}{a} \qquad (8.9)$$

because the sense of rotation, and therefore the orientation of the hand (and thumb), has to be reversed to go from $\underset{\sim}{b}$ to $\underset{\sim}{a}$. Also, since $\sin\theta = 0$ if $\theta = 0$, the cross product will be zero if two vectors are parallel to each other.

To evaluate the vector product of $\underset{\sim}{a}$ and $\underset{\sim}{b}$ in terms of their Cartesian components, we first need to note the results of taking cross products between their basis vectors: $\underset{\sim}{i} \times \underset{\sim}{i} = \underset{\sim}{j} \times \underset{\sim}{j} = \underset{\sim}{k} \times \underset{\sim}{k} = 0$, and $\underset{\sim}{i} \times \underset{\sim}{j} = \underset{\sim}{k}$, $\underset{\sim}{j} \times \underset{\sim}{i} = -\underset{\sim}{k}$, and so on. Then, expanding out the cross product of $a_1 \underset{\sim}{i} + a_2 \underset{\sim}{j} + a_3 \underset{\sim}{k}$ with $b_1 \underset{\sim}{i} + b_2 \underset{\sim}{j} + b_3 \underset{\sim}{k}$, we find that $\underset{\sim}{a} \times \underset{\sim}{b}$ reduces to

$$\underset{\sim}{a} \times \underset{\sim}{b} = (a_2 b_3 - b_3 a_2, a_3 b_1 - b_3 a_1, a_1 b_2 - b_1 a_2) \qquad (8.10)$$

and we'll see a succinct way of remembering this in section 9.3.

Geometrically, the magnitude of a cross product is the area of a parallelogram, or twice the area of the triangle, whose adjacent sides are given by $\underset{\sim}{a}$ and $\underset{\sim}{b}$. Indeed, by definition, the *vector area* of the parallelogram is given by $\underset{\sim}{a} \times \underset{\sim}{b}$, with the direction giving a unique normal to the planar surface.

Area of triangle $= |\underset{\sim}{a} \times \underset{\sim}{b}|/2$

To see the cross product in action we can post-multiply both sides of eqn (8.2) by $(\underset{\sim}{b} - \underset{\sim}{a})$ and, noting that $(\underset{\sim}{b} - \underset{\sim}{a}) \times (\underset{\sim}{b} - \underset{\sim}{a}) = 0$, eliminate the scalar parameter λ to obtain

$$\underset{\sim}{r} \times (\underset{\sim}{b} - \underset{\sim}{a}) = \underset{\sim}{a} \times (\underset{\sim}{b} - \underset{\sim}{a}) = \underset{\sim}{a} \times \underset{\sim}{b} \qquad (8.11)$$

which is an alternative form of the vector equation of a line.

8.5 Scalar triple product

We next explore the concept of 'multiplying three vectors', $\underset{\sim}{a}$, $\underset{\sim}{b}$, and $\underset{\sim}{c}$. The easiest way of doing this is the *scalar triple product* which is written $\underset{\sim}{a} \cdot (\underset{\sim}{b} \times \underset{\sim}{c})$, or sometimes $[\underset{\sim}{a}, \underset{\sim}{b}, \underset{\sim}{c}]$, and produces a scalar. In component form this gives

$$\underset{\sim}{a} \cdot (\underset{\sim}{b} \times \underset{\sim}{c}) = a_1(b_2 c_3 - b_3 c_2) + a_2(b_3 c_1 - b_1 c_3) + a_3(b_1 c_2 - b_2 c_1) \qquad (8.12)$$

where again a more compact notation will be introduced in section 9.3.

The scalar triple product measures the volume of the *parallelepiped* whose edges are formed by the vectors $\underset{\sim}{a}$, $\underset{\sim}{b}$ and $\underset{\sim}{c}$. We saw in section 8.4 that $\underset{\sim}{b} \times \underset{\sim}{c}$ gives the area of the parallelogram with sides $\underset{\sim}{b}$ and $\underset{\sim}{c}$. The dot product with $\underset{\sim}{a}$ multiplies this area by the perpendicular height, given by the projection of $\underset{\sim}{a}$ along the normal to the base parallelogram, and thus generates the volume. Since volume is a fixed physical quantity, the magnitude of the scalar triple product is independent of the order of $\underset{\sim}{a}$, $\underset{\sim}{b}$ and $\underset{\sim}{c}$, but we must take some

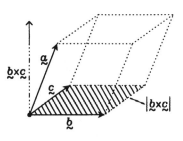

Volume of tetrahedron
$$= |\underset{\sim}{a} \cdot (\underset{\sim}{b} \times \underset{\sim}{c})|/6$$

care with its sign. Specifically, the *cyclic combinations* of $\underset{\sim}{a}$, $\underset{\sim}{b}$ and $\underset{\sim}{c}$ are all equal: $\underset{\sim}{a} \cdot (\underset{\sim}{b} \times \underset{\sim}{c}) = \underset{\sim}{b} \cdot (\underset{\sim}{c} \times \underset{\sim}{a}) = \underset{\sim}{c} \cdot (\underset{\sim}{a} \times \underset{\sim}{b})$. The *ant-cyclic* ones, $\underset{\sim}{a} \cdot (\underset{\sim}{c} \times \underset{\sim}{b}) = \underset{\sim}{b} \cdot (\underset{\sim}{a} \times \underset{\sim}{c}) = \underset{\sim}{c} \cdot (\underset{\sim}{b} \times \underset{\sim}{a})$, are all equal to $-\underset{\sim}{a} \cdot (\underset{\sim}{b} \times \underset{\sim}{c})$, as can be seen most clearly using eqn (8.9).

If any two vectors in a scalar triple product are parallel, or the same, then its value must be zero because it is the volume of a parallelepiped of zero height. This provides a convenient test of whether three vectors, $\underset{\sim}{a}$, $\underset{\sim}{b}$, and $\underset{\sim}{c}$, lie in a plane, and are *co-planar*: if this is the case then the scalar triple product $\underset{\sim}{a} \cdot (\underset{\sim}{b} \times \underset{\sim}{c})$ is necessarily zero. Three non-coplanar vectors are said to span a three-dimensional vector space in the sense that any point in it can be expressed in the form

$$\underset{\sim}{r} = l\underset{\sim}{a} + m\underset{\sim}{b} + n\underset{\sim}{c} \tag{8.13}$$

where l, m, and n are scalars. If two of the three vectors are parallel, then the vectors only span a two-dimensional space and are said to be *linearly dependent* in that we can always find scalars p and q such that

$$\underset{\sim}{c} = p\underset{\sim}{a} + q\underset{\sim}{b} \tag{8.14}$$

To find l in eqn (8.13), for example, we take the scalar product of both sides with $\underset{\sim}{b} \times \underset{\sim}{c}$, generating a scalar equation, and then divide both sides by the scalar $\underset{\sim}{a} \cdot (\underset{\sim}{b} \times \underset{\sim}{c})$.

To end this section let us use the scalar triple product to derive an alternative form for the vector equation of a plane given by eqn (8.3). First, take the scalar product of both sides of this equation with the vector

$$\underset{\sim}{n} = (\underset{\sim}{b} - \underset{\sim}{a}) \times (\underset{\sim}{c} - \underset{\sim}{a}) = \underset{\sim}{b} \times \underset{\sim}{c} + \underset{\sim}{a} \times \underset{\sim}{b} + \underset{\sim}{c} \times \underset{\sim}{a}$$

noting it makes no odds whether we pre- or post-multiply. We choose this form of $\underset{\sim}{n}$ to ensure that both of the arbitrary constants, μ and ν, are eliminated: this occurs because they are multiplied by scalar triple products which, because they contain two parallel vectors, have values of zero. The direction of this vector is normal to the plane containing the vectors $(\underset{\sim}{b} - \underset{\sim}{a})$ and $(\underset{\sim}{c} - \underset{\sim}{a})$. We obtain

$$\frac{\underset{\sim}{r} \cdot \underset{\sim}{n}}{|\underset{\sim}{n}|} = \frac{\underset{\sim}{a} \cdot (\underset{\sim}{b} \times \underset{\sim}{c})}{|\underset{\sim}{n}|} = D \tag{8.15}$$

where we have divided both sides by $|\underset{\sim}{n}|$ to make the left-hand side the scalar product of $\underset{\sim}{r}$ with a unit vector.pr This provides a clear picture of the geometry of a plane: the left-hand side of eqn (8.15) is the projection of $\underset{\sim}{r}$ along the unit normal to the plane, so that D is the perpendicular distance from the origin to the plane. The Cartesian equation of the plane can be obtained by substituting $\underset{\sim}{r} = (x, y, z)$ into eqn (8.15).

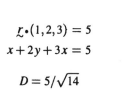

$$\underset{\sim}{r} \cdot (1, 2, 3) = 5$$

$$x + 2y + 3x = 5$$

$$D = 5/\sqrt{14}$$

8.6   Vector triple product

Another way of combining triplets of vectors is with the *vector triple product*, written $\underset{\sim}{a} \times (\underset{\sim}{b} \times \underset{\sim}{c})$, which produces a vector. It can be simplified using

$$\underset{\sim}{a} \times (\underset{\sim}{b} \times \underset{\sim}{c}) = \underbrace{(\underset{\sim}{a} \cdot \underset{\sim}{c}) \underset{\sim}{b}}_{AB} - \underbrace{(\underset{\sim}{a} \cdot \underset{\sim}{b}) \underset{\sim}{c}}_{AB} \tag{8.16}$$

which can be remembered as the 'ABACAB' identity. An example of the use of eqn (8.16) is given by one method of finding the line of intersection of two planes. Following eqn (8.15) we can write these two planes as

$$\underset{\sim}{r} \cdot \underset{\sim}{a} = u \quad \text{and} \quad \underset{\sim}{r} \cdot \underset{\sim}{b} = v$$

If we now consider the vector triple product, $\underset{\sim}{r} \times (\underset{\sim}{a} \times \underset{\sim}{b})$, and apply the 'ABACAB' identity, we obtain

$$\underset{\sim}{r} \times (\underset{\sim}{a} \times \underset{\sim}{b}) = (\underset{\sim}{r} \cdot \underset{\sim}{b}) \underset{\sim}{a} - (\underset{\sim}{r} \cdot \underset{\sim}{a}) \underset{\sim}{b} = v\underset{\sim}{a} - u\underset{\sim}{b}$$

where we have used the fact that $\underset{\sim}{r}$ must lie on both planes. A comparison with eqn (8.11) shows that we have effortlessly recovered the vector equation of the line of intersection, the direction being given by $(\underset{\sim}{a} \times \underset{\sim}{b})$.

8.7 Polar coordinates

The location of a point on a graph, or in two-dimensional space, is usually specified with Cartesian coordinates (x, y). In section 7.3, however, we noted that an alternative way of defining the position was in terms of the distance from the origin, r, and the anticlockwise angle, θ, made by this 'radius' with the positive x-axis; the latter are known as *polar* coordinates, and are written as (r, θ). The relationship between the Cartesian and polar forms is given in eqns (7.9) and (7.10).

$$x = r \cos\theta$$
$$y = r \sin\theta$$
$$r^2 = x^2 + y^2$$
$$\theta = \tan^{-1}(y/x)$$

When working in three dimensions, there are two commonly-used generalisations of the polar formulation. The first is very easy: x and y are replaced with r and θ, as above, but z is kept unchanged. Thus the triplet (r, θ, z) constitutes *cylindrical polar* coordinates. In the second approach, the location of a point is specified by its distance from the origin, or the radius r, and two angles, θ and ϕ, which correspond to its co-latitude and longitude. That is to say, θ is measured from the z-axis so that $0°$ is 'due north', $90°$ is in the x–y plane (with $z = 0$) and $180°$ is in the direction of the 'south pole'; the longitude ϕ is the anticlockwise angle from the positive x-axis to the projection of the radius on to the 'equatorial' plane. Following the global overtones, (r, θ, ϕ) are called *spherical polar* coordinates. A bit of thought, and elementary trigonometry, shows that r, θ and ϕ are related to their Cartesian counterparts by

$$\begin{aligned} x &= r \sin\theta \cos\phi \\ y &= r \sin\theta \sin\phi \\ z &= r \cos\theta \end{aligned} \quad (8.17)$$

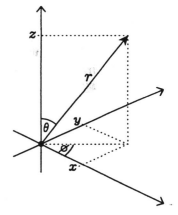

The reverse transformations can also be ascertained, of which the most useful is the familiar result that $r^2 = x^2 + y^2 + z^2$.

Although it may seem unnecessary to have different coordinate systems when a Cartesian choice will suffice, and is so much easier to manipulate, we have already seen the value of the polar formulation in chapter 7. The advantages of working in a reference system which matches the natural geometry of the problem at hand, such as a simplification of the related algebra, will also be met in chapter 12. While we have only considered the cases relevant for rectangular, cylindrical and spherical situations here, there are many alternatives; in practice, they are rarely used.

Exercises

8.1 State which of the following quantities can be described by vectors:
(a) temperature, (b) acceleration, (c) force, (d) molecular weight,
(e) area, (f) internal energy, (g) magnetic field.

8.2 The points A, B, C, and D have position vectors $a = (1,2,3)$, $b = (2,0,1)$,
$c = (1,1,1)$, and $d = (5,2,5)$ respectively. Calculate: (i) $a + b + c + d$,
(ii) $a + b - c - d$, (iii) $2a - 3b - 5c + d/2$, (iv) $a + 2b + 0c - d$, (v) the
position vectors of the mid-points of $\vec{BC}$ and $\vec{AD}$.

8.3 Taking a, b, c and d from exercise 8.2 find: (i) the vector equation of the
line passing through A and C, (ii) the vector equation passing through the
mid-points of $\vec{AB}$ and $\vec{CD}$, and (iii) the Cartesian equations of the lines
defined by $r = a + \lambda b$ and $r = c + \lambda d$.

8.4 Taking a, b, c and d from exercise 8.2 find $a \cdot b$, $a \cdot c$, and $a \cdot d$. Find
the angles between b and c, and between c and d. Evaluate $(a \cdot c) b$ and
$(a \cdot b) c$.

8.5 Use the dot product to obtain the identity of eqn (3.13).

8.6 Taking a, b, c and d from exercise 8.2 find: $a \times b$, $a \times c$, and $a \times d$.
Find the angles between b and c, and between c and d. Express the two
lines discussed in exercise 8.3(iii) in the form described by eqn (8.11).

8.7 Use the cross product to obtain the 'sine rule' of eqn (3.22).

8.8 Taking a, b, c and d from exercise 8.2 find $a \cdot (b \times c)$, $a \cdot (c \times d)$ and
$a \cdot (b \times d)$. Use these to determine the three position vectors which are
coplanar, and find an equation for this plane in its Cartesian form, and
in the form of eqn (8.15). What is the perpendicular distance from the
origin to this plane?

8.9 Find the scalar triple product of the vectors $(1,2,4)$, $(2,0,-3)$, $(-4,4,17)$.
Are they linearly independent? Can the third vector be expressed as a
linear combination of the first two; if so, find this linear combination.

8.10 Use the results of exercise 8.4 to verify the rule of eqn (8.16).

8.11 The *reciprocal vectors* of the set a, b and c are defined by

$$a' = b \times c/s$$
$$b' = c \times a/s$$
$$c' = a \times b/s$$

where $s = a \cdot (b \times c)$. Show that $a' \cdot a = b' \cdot b = c' \cdot c = 1$, that $a' \cdot b = a' \cdot c = 0$, and evaluate the scalar triple product of the reciprocal vectors
in terms of s. If a given vector x is expressed as a linear combination of
these reciprocal vectors, show that the coefficient of a' is $a \cdot x$.

9 Matrices

9.1 Definition and nomenclature

In this chapter we will be considering the topic of *matrices*. Although our discussion may seem rather abstract at the outset, it is an important topic that often goes under the heading of *linear algebra*.

The simplest definition of a matrix is that it is a rectangular array of numbers, usually enclosed in (large) round brackets. If it consists of M rows and N columns, it is said to be an $M \times N$ matrix. For example

$$\underset{\approx}{A} = \begin{pmatrix} 2 & 5 & 3 \\ 1 & 4 & 7 \end{pmatrix} \tag{9.1}$$

is a 2×3 matrix. If the matrix is denoted by the symbol $\underset{\approx}{A}$, as in eqn (9.1), then its individual *elements* are specified by appending two suffices, i and j, to A that give their locations in terms of row and column respectively: A_{ij}, where $i = 1$ or 2 and $j = 1, 2$, or 3 in the case of eqn (9.1). Thus $A_{11} = 2, A_{21} = 1, A_{12} = 5$, and so on.

$$\underset{\approx}{A} = \begin{pmatrix} A_{11} & A_{12} & A_{13} & \cdots & A_{1N} \\ A_{21} & A_{22} & A_{23} & \cdots & A_{2N} \\ \vdots & \vdots & \vdots & \vdots & \vdots \\ A_{M1} & A_{M2} & A_{M3} & \cdots & A_{MN} \end{pmatrix}$$

Depending on their size (or shape), and the properties of their elements, matrices are often given qualifying names. For example, a *row* matrix has only one row ($M = 1$) and a *column* matrix consists of a single column ($N = 1$); in fact these can be considered to be vectors, requiring just one suffix, with the elements being the relevant components. If the number of rows and columns are equal ($M = N$), then it is said to be a *square* matrix; this case is of special interest, and the bulk of this chapter is devoted to its study.

Two properties of matrices which are frequently met in practice is that they are *real* and *symmetric*. The former means the elements are not complex numbers and the second, which implicitly assumes a square matrix, that the matrix remains the same if its rows and columns are interchanged. The operation of switching around rows and columns is called a *transpose*, and is denoted by attaching a 'T' superscript to the matrix; thus we can formally write that the ij^{th} element of $\underset{\approx}{A}^T$ is equal to the ji^{th} element of $\underset{\approx}{A}$

$$\underset{\approx}{A}^T = \begin{pmatrix} A_{11} & A_{21} & \cdots & A_{M1} \\ A_{12} & A_{22} & \cdots & A_{M2} \\ A_{13} & A_{23} & \cdots & A_{M3} \\ \vdots & \vdots & \vdots & \vdots \\ A_{1N} & A_{2N} & \cdots & A_{MN} \end{pmatrix}$$

$$\left(A^T\right)_{ij} = A_{ji} \tag{9.2}$$

The transpose of a column matrix is, of course, a row matrix, and vice versa. A real and symmetric matrix is, therefore, defined by the properties that

$$\underset{\approx}{A}^* = \underset{\approx}{A} \quad \text{and} \quad \underset{\approx}{A}^T = \underset{\approx}{A} \tag{9.3}$$

where the superscripted '*' corresponds to the operation of replacing each element by its complex conjugate. A type of matrix which plays a central rôle in quantum mechanics is called *hermitian*; it satisfies the condition that

$$\underset{\approx}{A}^T = \underset{\approx}{A}^* \tag{9.4}$$

A consideration of eqns (9.3) and (9.4) shows that a real and symmetric matrix is a special case of a hermitian one.

9.2   Matrix arithmetic

The sum or difference of two matrices, say $\underset{\approx}{A}$ and $\underset{\approx}{B}$, is calculated by adding or subtracting the corresponding elements

$$\underset{\approx}{C} = \underset{\approx}{A} \pm \underset{\approx}{B} \quad \Longleftrightarrow \quad C_{ij} = A_{ij} \pm B_{ij} \tag{9.5}$$

$$\begin{pmatrix} 2 & 5 \\ 1 & 4 \\ 6 & 1 \end{pmatrix} + \begin{pmatrix} 3 & -1 \\ 7 & 0 \\ -3 & 1 \end{pmatrix} = \begin{pmatrix} 5 & 4 \\ 8 & 4 \\ 3 & 2 \end{pmatrix}$$

Both matrices must be of the same size for this combination to be viable; in other words, if $\underset{\approx}{A}$ is an $N \times M$ matrix so too must be $\underset{\approx}{B}$ (and $\underset{\approx}{C}$).

Matrix multiplication is far less straightforward than addition and subtraction. It is defined as follows: the ij^{th} element of the product of $\underset{\approx}{A}$ and $\underset{\approx}{B}$ is given by the scalar, or dot, product of the ith row of $\underset{\approx}{A}$ with the jth column of $\underset{\approx}{B}$. Explicitly

$$\underset{\approx}{C} = \underset{\approx}{A}\, \underset{\approx}{B} \quad \Longleftrightarrow \quad C_{ij} = \sum_k A_{ik} B_{kj} \tag{9.6}$$

$$\begin{pmatrix} 2 & 5 \\ 1 & 4 \\ 6 & 1 \end{pmatrix} \begin{pmatrix} 3 & -1 \\ 7 & 0 \end{pmatrix} = \begin{pmatrix} 41 & -2 \\ 31 & -1 \\ 25 & -6 \end{pmatrix}$$

where the summation over the index k requires that $\underset{\approx}{A}$ has the same number of columns as $\underset{\approx}{B}$ has rows. The usefulness of the somewhat perverse rule of eqn (9.6) will become clearer later, but we should note that it leads to an algebra where multiplication is no longer commutative; that is to say

$$\underset{\approx}{A}\,\underset{\approx}{B} \neq \underset{\approx}{B}\,\underset{\approx}{A}$$

in general. Indeed, $\underset{\approx}{B}\,\underset{\approx}{A}$ may not even exist, even if $\underset{\approx}{A}\,\underset{\approx}{B}$ does, because of the size restrictions on the numbers of rows and columns needed for a product. This lack of commutativity means that we must be very careful in stating whether both sides of an equation are to be pre- or post-multiplied by a matrix — the order matters!

Incidentally, the transpose of a product is equal to the product of the transposes but in reverse order

$$\left(\underset{\approx}{A}\,\underset{\approx}{B} \right)^T = \underset{\approx}{B}^T \underset{\approx}{A}^T \tag{9.7}$$

a result which can be derived from eqns (9.2) and (9.6).

$$2 \begin{pmatrix} 3 & -1 \\ 7 & 0 \\ -3 & 1 \end{pmatrix} = \begin{pmatrix} 6 & -2 \\ 14 & 0 \\ -6 & 2 \end{pmatrix}$$

Division is not allowed in matrix algebra, but we will shortly see how this limitation can sometimes be circumvented. The only exception is division by a scalar (or a 1×1 matrix) which, like its multiplication counterpart, is an anomalous case. It is a very simple operation

$$\underset{\approx}{C} = \alpha \underset{\approx}{B} \quad \Longleftrightarrow \quad C_{ij} = \alpha\, B_{ij} \tag{9.8}$$

where every element of the matrix is multiplied or divided by the constant α.

9.3   Determinants

Before continuing with our general study of matrices, we must make a brief digression to consider the topic of *determinants*. This is an essential prerequisite for much of the discussion about square matrices that is to follow in the rest of this chapter.

The easiest case is that of a 1×1 matrix, whose determinant is simply equal to the element itself: $\det(\underset{\approx}{A}) = |A_{11}| = A_{11}$. Moving onto a less trivial situation, the determinant of a 2×2 matrix is defined to be

$$\det(\underset{\approx}{A}) = \begin{vmatrix} A_{11} & A_{12} \\ A_{21} & A_{22} \end{vmatrix} = A_{11}A_{22} - A_{12}A_{21} \qquad (9.9)$$

The explicit formulae for the determinants of higher-order matrices become increasingly complicated, but can always be related to ones of a lower-order through a straightforward rule: the determinant of a matrix is given by the scalar, or dot, product of any of its rows or columns with its corresponding *cofactors*. The cofactor of the matrix-element A_{ij} is, in turn, given by $(-1)^{i+j}$ times the determinant of the matrix which remains when the ith row and j^{th} column of $\underset{\approx}{A}$ are removed; the former yields a checker-board scheme of $+1$ and -1, and the latter, known as the *minor*, is the determinant of a smaller matrix (by one) than the original.

Thus, for example, the determinant of a 3×3 matrix can be written in terms of the determinants of three 2×2 matrices

$$\begin{vmatrix} A_{11} & A_{12} & A_{13} \\ A_{21} & A_{22} & A_{23} \\ A_{31} & A_{32} & A_{33} \end{vmatrix} = A_{11}\begin{vmatrix} A_{22} & A_{23} \\ A_{32} & A_{33} \end{vmatrix} - A_{12}\begin{vmatrix} A_{21} & A_{23} \\ A_{31} & A_{33} \end{vmatrix} + A_{13}\begin{vmatrix} A_{21} & A_{22} \\ A_{31} & A_{32} \end{vmatrix}$$

where we have carried out the calculation with respect to the top row. Although unnecessary given eqn (9.9), the determinant of a 2×2 matrix can itself be written in terms of the determinants of two 1×1 matrices; this is a useful exercise to carry out (once), since it enables us to verify that the determinant yielded by the cofactor rule is independent of the choice of row or column.

Determinants have a certain number of general properties, some of which are useful in facilitating their evaluation. These are listed without proof, but can be made plausible by checking that they work explicitly for a 2×2 matrix: (i) interchanging rows with columns leaves the determinant unchanged; (ii) multiplying a row or column by a constant k multiplies the determinant by k; (iii) if one row (or column) is zero then, then so too is the determinant; (iv) if one row (or column) is a multiple of another, then the determinant is zero; (v) interchanging two adjacent rows (or columns) multiplies the determinant by -1; (vi) the addition of a multiple of one row (or column) to another leaves the determinant unchanged; (vii) the determinant of a product of matrices is equal to the product of their determinants.

Incidentally, the cofactor rule for determinants provides a good method for remembering the formula for a vector, or cross, product met in section 8.4

$$\underset{\sim}{a} \times \underset{\sim}{b} = \begin{vmatrix} \underset{\sim}{i} & \underset{\sim}{j} & \underset{\sim}{k} \\ a_x & a_y & a_z \\ b_x & b_y & b_z \end{vmatrix} \qquad (9.10)$$

What's more, it's not too difficult to see that a scalar triple product is obtained if the top row is replaced by the components of a third vector $\underset{\sim}{c}$. Thus, with reference to section 8.5, the determinant of a 3×3 matrix represents the volume of a parallelepiped. In fact, this is the general physical interpretation of a determinant — its magnitude gives the 'volume' of the object generated when rows,

$$\begin{vmatrix} 2 & 5 \\ 1 & 4 \end{vmatrix} = 2 \times 4 - 5 \times 1 = 3$$

$$\begin{pmatrix} + & - & + & \cdots & \cdots \\ - & + & - & \cdots & \cdots \\ + & - & + & \cdots & \cdots \\ \vdots & \vdots & \vdots & \vdots & + \end{pmatrix}$$

$$\begin{vmatrix} 1 & 0 & 2 \\ 3 & -1 & 5 \\ 4 & 3 & 7 \end{vmatrix} = \begin{vmatrix} -1 & 5 \\ 3 & 7 \end{vmatrix} + 2\begin{vmatrix} 3 & -1 \\ 4 & 3 \end{vmatrix}$$
$$= -7 - 15 + 2(9 + 4) = 4$$

$$\det(\underset{\approx}{A}^T) = \det(\underset{\approx}{A})$$

$$\det(\underset{\approx}{A}\,\underset{\approx}{B}) = \det(\underset{\approx}{A})\det(\underset{\approx}{B})$$

$$(\underset{\sim}{a} \times \underset{\sim}{b}) \cdot \underset{\sim}{c} = \begin{vmatrix} c_x & c_y & c_z \\ a_x & a_y & a_z \\ b_x & b_y & b_z \end{vmatrix}$$

or columns, are used as bounding vectors. This reduces to a simple length for the case of a 1×1 matrix, and the area of a parallelogram for a 2×2 one.

9.4   Inverse matrices

We said in section 9.2 that division by a matrix was not allowed. There is a manipulation for square matrices, however, that mimics this operation: multiplication by an *inverse* matrix. The latter is defined by the property that

$$\underset{\approx}{A}^{-1}\underset{\approx}{A} = \underset{\approx}{A}\,\underset{\approx}{A}^{-1} = \underset{\approx}{I} \tag{9.11}$$

$$\underset{\approx}{I} = \begin{pmatrix} 1 & 0 & 0 & \dots & 0 \\ 0 & 1 & 0 & \dots & 0 \\ 0 & 0 & 1 & \dots & 0 \\ \vdots & \vdots & \vdots & \ddots & \vdots \\ 0 & 0 & 0 & \dots & 1 \end{pmatrix}$$

where $\underset{\approx}{I}$ is the matrix equivalent of unity, consisting of ones down the diagonal and zeros everywhere else, and is called an *identity* or *unit* matrix; anything (permissible) multiplied by $\underset{\approx}{I}$ returns the entity itself.

Although the inverse of a matrix could be ascertained from the definition of eqn (9.11), a more systematic approach is provided by the following result

$$\underset{\approx}{A}^{-1} = \frac{\mathrm{adj}\left(\underset{\approx}{A}\right)}{\det\left(\underset{\approx}{A}\right)} \tag{9.12}$$

$$\begin{pmatrix} 5 & 3 \\ 2 & 1 \end{pmatrix}^{-1} = \frac{1}{5-6}\begin{pmatrix} 1 & -3 \\ -2 & 5 \end{pmatrix}$$

$$= \begin{pmatrix} -1 & 3 \\ 2 & -5 \end{pmatrix}$$

where $\mathrm{adj}(\underset{\approx}{A})$ is the *adjoint* of $\underset{\approx}{A}$; it is a matrix which is constructed by replacing each of the elements of the transpose of $\underset{\approx}{A}$, $\underset{\approx}{A}^T$, by its cofactors. The denominator in eqn (9.12) indicates that the inverse of $\underset{\approx}{A}$, $\underset{\approx}{A}^{-1}$, does not exist if its determinant, $\det(\underset{\approx}{A})$, is equal to zero; such a matrix is said to be *singular*.

9.5   Linear simultaneous equations

We met simultaneous equations in section 1.4, and noted that the easiest ones to solve are those which are linear. In the context of the current chapter, these are ones that can be cast in the following general terms

$$\underset{\approx}{A}\,\underset{\sim}{X} = \underset{\sim}{B} \tag{9.13}$$

where $\underset{\approx}{A}$ is a matrix, and $\underset{\sim}{X}$ and $\underset{\sim}{B}$ are vectors or column matrices; the elements of $\underset{\approx}{A}$ and $\underset{\sim}{B}$ are known whereas those of $\underset{\sim}{X}$, which satisfy eqn (9.13), are sought. For example, the pair of equations in section 1.4 can be written as

$$\begin{pmatrix} a & b \\ c & d \end{pmatrix}\begin{pmatrix} x \\ y \end{pmatrix} = \begin{pmatrix} \alpha \\ \beta \end{pmatrix}$$

where a, b, c, d, α and β are constants.

A formula for the solution of eqn (9.13) is readily obtained if both sides are premultiplied by the inverse $\underset{\approx}{A}^{-1}$

$$\underset{\approx}{A}^{-1}\underset{\approx}{A}\,\underset{\sim}{X} = \underset{\approx}{A}^{-1}\underset{\sim}{B} \quad \Rightarrow \quad \underset{\approx}{I}\,\underset{\sim}{X} = \underset{\sim}{X} = \underset{\approx}{A}^{-1}\underset{\sim}{B} \tag{9.14}$$

$$\begin{pmatrix} x \\ y \end{pmatrix} = \begin{pmatrix} a & b \\ c & d \end{pmatrix}^{-1}\begin{pmatrix} \alpha \\ \beta \end{pmatrix}$$

$$= \frac{1}{\Delta}\begin{pmatrix} d & -b \\ -c & a \end{pmatrix}\begin{pmatrix} \alpha \\ \beta \end{pmatrix}$$

for $\Delta = ad - bc \neq 0$

where we have used the property of eqn (9.11), and that of the identity matrix. The evaluation of the inverse may require some effort, of course, but the actual prescription of eqn (9.14) for finding $\underset{\sim}{X}$ is straightforward.

From our discussion in section 9.4, it is clear that we have implicitly assumed that: (i) the matrix $\underset{\approx}{A}$ is square; and (ii) $\det(\underset{\approx}{A}) \neq 0$. The first point means that there must be as many simultaneous equations as unknowns, and the second ensures that they are *linearly independent*. That is to say we can't construct any one of them from a simple combination of the others, so that there are really fewer equations than it would appear. If $\underset{\approx}{A}$ is singular, then either the solution is not unique or there is no solution at all. Geometrically, for the case of a 2×2 matrix, we seek the intersection of two straight lines. They will not meet at a single point if they are parallel; if they lie on top of each other, then everywhere along the line is a solution; otherwise, the simultaneous conditions are never satisfied.

9.6 Transformations

When a matrix (not necessarily square) is applied to a vector, it generates a new column matrix (or vector)

$$\underset{\approx}{A}\,\underset{\sim}{X} = \underset{\sim}{Y} \qquad (9.15)$$

The change from the 'input' $\underset{\sim}{X}$ to the 'output' $\underset{\sim}{Y}$ is called a *mapping*, or a *transformation*, with the matrix $\underset{\approx}{A}$ often referred to as an *operator*. For example, when a two-dimensional vector is premultiplied by

$$\underset{\approx}{A} = \begin{pmatrix} \cos\theta & -\sin\theta \\ \sin\theta & \cos\theta \end{pmatrix} \qquad (9.16)$$

it rotates it anticlockwise through an angle θ. As with all 2×2 matrices, the nature of the transformation is best ascertained by considering its effect on the corners of the unit square. That is, by working out where the following points map to: $(0,0)$, $(0,1)$, $(1,0)$ and $(1,1)$. If $\underset{\approx}{A}$ is square, and not singular, then a reverse transformation is obtained by applying the inverse, $\underset{\approx}{A}^{-1}$, to $\underset{\sim}{Y}$.

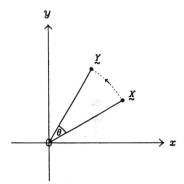

9.7 Eigenvalues and eigenvectors

There is a special case of eqn (9.15), called the *eigenvalue* equation, which is of great physical importance. That is when $\underset{\sim}{X}$ maps onto itself, give or take a scale factor λ

$$\underset{\approx}{A}\,\underset{\sim}{X} = \lambda\,\underset{\sim}{X} \qquad (9.17)$$

The values of λ and $\underset{\sim}{X}$ which satisfy this equation are referred to as the eigenvalues and *eigenvectors* of $\underset{\approx}{A}$ respectively.

The matrix $\underset{\approx}{A}$ here is implicitly square, so eqn (9.17) becomes

$$\left(\underset{\approx}{A} - \lambda\,\underset{\approx}{I}\right)\underset{\sim}{X} = 0 \qquad (9.18)$$

where we have made use of the property that $\underset{\approx}{I}\,\underset{\sim}{X} = \underset{\sim}{X}$, and the unit matrix is of the same size as $\underset{\approx}{A}$. Following our discussion in section 9.5, premultiplication of both sides of eqn (9.18) by the inverse of the composite matrix $\left(\underset{\approx}{A} - \lambda\,\underset{\approx}{I}\right)$ leads to the conclusion that $\underset{\sim}{X} = 0$. The only way we can escape this fate is if

$$\det\left(\underset{\approx}{A} - \lambda\,\underset{\approx}{I}\right) = 0 \qquad (9.19)$$

for then $\left(\underset{\approx}{A} - \lambda\,\underset{\approx}{I}\right)^{-1}$ does not exist, and so we are not forced to accept the trivial solution of $\underset{\sim}{X} = 0$.

The use of eqn (9.19) is usually the first step in solving an eigenvalue equation. It tells us to subtract λ from each of the diagonal elements of $\underset{\approx}{A}$, and to set the determinant of the resulting matrix equal to zero. For an $N \times N$ matrix, this gives rise to the problem of finding the roots of an Nth-order poly-

$$\begin{vmatrix} 4-\lambda & 1 \\ 2 & 3-\lambda \end{vmatrix} = 0$$

$$\Rightarrow \quad \lambda^2 - 7\lambda + 10 = 0$$

$$\Rightarrow \quad \lambda = 2 \quad \text{or} \quad \lambda = 5$$

nomial in λ (called the *characteristic* equation). Thus there are N eigenvalues $(\lambda_1, \lambda_2, \lambda_3, \ldots, \lambda_N)$, and we have to consider each in turn to find the corresponding eigenvectors $(\underset{\sim}{X}_1, \underset{\sim}{X}_2, \underset{\sim}{X}_3, \ldots, \underset{\sim}{X}_N)$

When $\lambda = 2$, $2x + y = 0$

$$\Rightarrow \quad \underset{\sim}{X} = t\begin{pmatrix} 1 \\ -2 \end{pmatrix}$$

The best way to work out the eigenvectors is to substitute for the λ, one at a time, into eqn (9.17) or eqn (9.18). Then let one of the components of $\underset{\sim}{X}$ be equal to a parameter like t, and express the others in terms of it. The presence of the unknown variable (t) simply reflects the fact that the solution for $\underset{\sim}{X}$ is not a unique point; since the relationship between the elements is fixed, however, we are able to ascertain a definite direction. By convention, each eigenvector is *normalised* by picking a value for t which makes its length unity. Sometimes the above procedure runs into difficulties, in which case it is usually worth repeating it by choosing a different component as the reference.

When $\lambda = 5$, $-x + y = 0$

$$\Rightarrow \quad \underset{\sim}{X} = t\begin{pmatrix} 1 \\ 1 \end{pmatrix}$$

When dealing with problems of physical interest, the matrices that are encountered tend to be real and symmetric or hermitian; these are defined by eqns (9.3) and (9.4) respectively. It can be shown that such matrices have very convenient and pleasant eigen-properties. Namely, that all the eigenvalues are real, and the eigenvectors are mutually orthogonal

$$\lambda_i = \lambda_i^* \qquad \text{and} \qquad \underset{\sim}{X}_i^T \underset{\sim}{X}_j = \underset{\sim}{X}_i \cdot \underset{\sim}{X}_j = 0 \quad \text{if } i \neq j \tag{9.20}$$

$$\det\left(\underset{\approx}{A}\right) = \lambda_1 \lambda_2 \lambda_3 \cdots \lambda_N$$

$$\text{trace}\left(\underset{\approx}{A}\right) = \lambda_1 + \lambda_2 + \cdots + \lambda_N$$

where the subscripts label the different solutions of eqn (9.17). What's more, the product of the eigenvalues turns out to be equal to $\det\left(\underset{\approx}{A}\right)$ and their sum is given by the sum of the diagonal elements (or *trace*) of $\underset{\approx}{A}$.

9.8 Diagonalisation

We can obtain a geometrical interpretation of eigenvalues and eigenvectors by considering the following scalar quantity

$$Q = \underset{\sim}{X}^T \underset{\approx}{A} \underset{\sim}{X} \tag{9.21}$$

which is called a *quadratic form*. If $\underset{\approx}{A}$ is a real and symmetric 2×2 matrix, for example, with both eigenvalues having the same sign, and the vector $\underset{\sim}{X}^T$ is the row matrix $(x\ y)$ then eqn (9.21) yields the formula for an ellipse. The resulting equation won't be quite as simple as eqn (2.10), in general, because the ellipse will be skew with respect to the x and y axes. In this picture, the eigenvectors are the directions of the principal axes and the eigenvalues are inversely proportional to the squares of the corresponding widths. Higher-order matrices give rise to multidimensional ellipsoids, but the principal axes relationship to the eigen-properties still holds.

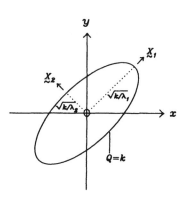

Having evaluated the eigenvalues and eigenvectors of $\underset{\approx}{A}$, any subsequent analysis can often be simplified by working in a new set of coordinates, $\underset{\sim}{Y}$, that are aligned with the principal axes of the ellipse. The required transformation between $\underset{\sim}{X}$ and $\underset{\sim}{Y}$ is given by

$$\underset{\sim}{X} = \underset{\approx}{Q} \underset{\sim}{Y} \tag{9.22}$$

$$\underset{\approx}{Q} = \begin{pmatrix} | & | & & | \\ \underset{\sim}{X}_1 & \underset{\sim}{X}_2 & \cdots & \underset{\sim}{X}_N \\ | & | & & | \end{pmatrix}$$

where the columns of the matrix $\underset{\approx}{Q}$ consist of the normalised eigenvectors of $\underset{\approx}{A}$; in conjunction with eqn (9.20), therefore, we find that the mapping matrix is *orthogonal* in that it satisfies the condition that

$$Q^T Q = I \tag{9.23}$$

With eqn (9.22), the quadratic form of eqn (9.21) reduces to

$$Q = Y^T \Lambda Y \tag{9.24}$$

where Λ is a *diagonal* matrix: it has zeros everywhere except down the diagonal, where the elements are equal to the eigenvalues of A.

The physical value of an eigen-analysis is that it decomposes apparently complicated behaviour into its basic constituent parts. The eigenvectors associated with molecular vibrations, for example, correspond to the *normal modes* of the system, and the eigenvalues give their natural frequencies. Similarly, the eigenvectors and eigenvalues encountered in a quantum mechanics problem relate to the wave functions of the *stationary states* and their energy levels.

Incidentally, if two eigenvalues are equal ($\lambda_1 = \lambda_2$, say) then the quadratic form is a circle; this is known as the *degenerate* case. Since there are no distinct principal axes, every direction in the plane is an eigenvector. As any two-dimensional vector can be constructed from two basis vectors, we are at liberty to select any two independent directions as being the eigenvectors; by convention, they are chosen to be orthogonal.

$$Q^T A Q = \Lambda$$

$$\Lambda = \begin{pmatrix} \lambda_1 & 0 & 0 & \cdots & 0 \\ 0 & \lambda_2 & 0 & \cdots & 0 \\ 0 & 0 & \lambda_3 & \cdots & 0 \\ \vdots & \vdots & \vdots & \ddots & \vdots \\ 0 & 0 & 0 & \cdots & \lambda_N \end{pmatrix}$$

Exercises

9.1 Find $A + B$, $A - B$, $A B$, and $B A$ when

$$A = \begin{pmatrix} 2 & 1 \\ 1 & 2 \end{pmatrix} \quad \text{and} \quad B = \begin{pmatrix} 3 & 3 \\ 0 & 4 \end{pmatrix}$$

Verify that $(A B)^T = B^T A^T$, and that $\det(A B) = \det(A)\det(B)$.

9.2 Confirm that eqn (9.10) gives the coordinates of a cross product, and that the determinant of a 3×3 matrix corresponds to a scalar triple product.

9.3 Find the inverse of the following matrix

$$C = \begin{pmatrix} 2 & -1 & 1 \\ 1 & -1 & 2 \\ -1 & 1 & -1 \end{pmatrix}$$

and verify that $C\,C^{-1} = C^{-1}C = I$.

9.4 Find the eigenvalues and eigenvectors of

$$A = \begin{pmatrix} 1 & 0 & 1 \\ 0 & -1 & 0 \\ 1 & 0 & 1 \end{pmatrix}$$

Confirm that the eigenvectors are mutually orthogonal, the sum of the eigenvalues is equal to the trace of A, and the product of eigenvalues is equal to $\det(A)$. Construct the diagonalisation matrix Q from the normalised eigenvectors of A, and verify that

$$Q\,Q^T = Q^T Q = I$$

Finally, check that the *similarity* transform $Q^T A\,Q = \Lambda$ yields a diagonal matrix simply related to the eigenvalues of A.

10 Partial differentiation

10.1 Definition and the gradient vector

In chapter 4, on differentiation, we were concerned with learning about the rate at which two quantities varied with respect to each other. In practice, of course, most problems involve several entities; for example, a thermodynamic variable, such as the internal energy of a gas, will depend on temperature, pressure and volume. The topic of *partial differentiation* is a natural extension of our earlier discussion.

To highlight the ideas clearly, and to avoid algebraic cluttering, we will focus principally on the case where a parameter z is a function of just two variables, x and y: $z = f(x,y)$. We can visualise such a situation from our everyday experience by thinking of z as representing a height, with x and y being the two-dimensional coordinates of a point on the (flat) ground. While a realistic rendering of the topographical scene requires artistic ability, a more practical means of displaying the salient information is to draw a *contour map*; these are found in geographical atlases, where hills and valleys are depicted by a series of lines which join together regions of equal altitude.

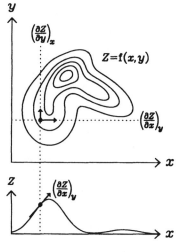

Given the aforementioned setup, how can we convey a sense of the slope at any given point (x,y)? Well, one possibility is to divide the problem into two parts where each is amenable to the analysis discussed in chapter 4. That is to say, we could take slices of the z-surface along the x and y directions which would give two sections, z against x (at fixed y) and z versus y (at constant x), that could be plotted as ordinary graphs. The pair of gradients which are generated in this way are called *partial derivatives*, because each provides some insight about the desired slope, and are denoted by curly d's

$$\left(\frac{\partial z}{\partial x}\right)_y \quad \text{and} \quad \left(\frac{\partial z}{\partial y}\right)_x$$

where the subscript specifies the quantity that is being held constant.

The procedure outlined above can be formalised mathematically as

$$\left(\frac{\partial z}{\partial x}\right)_y = \lim_{\delta x \to 0} \left[\frac{f(x+\delta x, y) - f(x,y)}{\delta x}\right]$$

$$\left(\frac{\partial z}{\partial y}\right)_x = \lim_{\delta y \to 0} \left[\frac{f(x, y+\delta y) - f(x,y)}{\delta y}\right] \tag{10.1}$$

where the formulae constitute the appropriate generalisation of of eqn (4.1). Although eqns (10.1) yield a 'first principles' derivation of the partial derivatives of $z = f(x,y)$, they are normally evaluated by appealing to the results in chapter 4 while treating the relevant parameter as a constant. For example, the derivative of $z = 3x^2 + y^3$ with respect to x at fixed y is easily seen to be $(\partial z/\partial x)_y = 6x$; similarly, it's fairly obvious that $(\partial z/\partial y)_x = 3y^2$. These can be confirmed with eqns (10.1)

$$\left(\frac{\partial z}{\partial x}\right)_y = \lim_{\delta x \to 0} \left[\frac{3(x+\delta x)^2 + y^3 - (3x^2+y^3)}{\delta x}\right] = \lim_{\delta x \to 0} \left[6x + 3\delta x\right]$$

and so on.

If we were standing on the side of a hill in real life, we wouldn't generally think of the partial derivatives as a means of talking about the local slope. Rather, we would tend to point up in the direction of the steepest ascent. This natural inclination is captured by the *gradient vector*, ∇z, defined by

$$\nabla z = \left(\frac{\partial z}{\partial x}, \frac{\partial z}{\partial y}\right) \tag{10.2}$$

where we have omitted the subscripts on the partial derivatives, as being implicitly given, for simplicity of notation. Thus the quantities discussed earlier in eqn (10.1) correspond to the x and y components of ∇z, a vector whose direction indicates the path of steepest ascent (or the greatest change in z) and whose magnitude gives the value of the slope along it. Since the gradient vector of eqn (10.2) is two-dimensional, it is best to think of it as being an arrow in the contour map; the properties of the latter, marking as it does the points of equal height (where there is no change in z), means that ∇z will be perpendicular to the contour lines.

Although we have only explicitly considered a function of two variables, the ideas generalise quite readily to multi-parameter problems. For example, if we were dealing with a function of three variables, $\Phi = f(x,y,z)$ say, then the gradient vector would be given by

$$\nabla \Phi = \left(\frac{\partial \Phi}{\partial x}, \frac{\partial \Phi}{\partial y}, \frac{\partial \Phi}{\partial z}\right) \tag{10.3}$$

where the relevant subscripts have again been omitted from the partial derivatives. Thus $\partial \Phi/\partial x$ should really read $(\partial \Phi/\partial x)_{yz}$, to give the derivative of Φ with respect to x at constant y and z; $\partial \Phi/\partial y$ should be $(\partial \Phi/\partial y)_{xz}$, and so on. The previous contour lines have to be replaced by three-dimensional surfaces upon which the value of Φ is fixed. The gradient vector at a point (x,y,z) is normal to the local 'contour surface' with its direction indicating the path of fastest increase in Φ; the magnitude of $\nabla \Phi$ gives the rate of greatest change.

10.2 Second and higher derivatives

Just as for ordinary derivatives in section 4.2, second and higher-order partial derivatives can also be evaluated. Thus with $z = f(x,y)$, for example,

$$\frac{\partial^2 z}{\partial x^2} = \frac{\partial}{\partial x}\left(\frac{\partial z}{\partial x}\right) \tag{10.4}$$

where $\partial/\partial x$ is the differential operator for the 'rate of change with respect to x, with y held fixed'; similarly, $\partial^2 z/\partial y^2 = \partial/\partial y(\partial z/\partial y)$. There are two other second partial derivatives but, some formal technicalities aside, these mixed terms are always equal

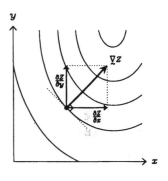

If $\Phi = 3x^2 + y^3 \sin x + \ln z$

$$\left(\frac{\partial \Phi}{\partial x}\right)_{yz} = 6x + y^3 \cos x$$

$$\left(\frac{\partial \Phi}{\partial y}\right)_{xz} = 3y^2 \sin x$$

$$\left(\frac{\partial \Phi}{\partial z}\right)_{xy} = \frac{1}{z}$$

$$\frac{\partial^2 \Phi}{\partial x^2} = 6 - y^3 \sin x$$

$$\frac{\partial^2 \Phi}{\partial y^2} = 6y \sin x$$

$$\frac{\partial}{\partial x}\left(\frac{\partial z}{\partial y}\right) = \frac{\partial}{\partial y}\left(\frac{\partial z}{\partial x}\right) \tag{10.5}$$

$$\frac{\partial^2 \Phi}{\partial x \partial y} = 3y^2 \cos x$$

and are denoted simply by $\partial^2 z/\partial x \partial y$. The relationship of eqn (10.5) is very important, and forms the basis of the derivation of many useful results.

10.3 Increments and chain rules

If $z = f(x, y)$, how much does the function change by when x and y are varied by a small amount? Well, the magnitude of the gradient vector tells us the rate of change of z in the direction of steepest ascent. The incremental change in z, δz, caused by a tiny move in the x–y plane is, therefore, given by the scalar, or dot, product of ∇z and the path vector $(\delta x, \delta y)$

$$\delta z \approx \left(\frac{\partial z}{\partial x}\right)_y \delta x + \left(\frac{\partial z}{\partial y}\right)_x \delta y \tag{10.6}$$

where equality is attained in the limit $\delta x \to 0$ and $\delta y \to 0$. The extension of eqn (10.6) to cases of three, or more, variables is straightforward; with $\Phi = f(x, y, z)$, say, we'll have

$$\delta \Phi \approx \left(\frac{\partial \Phi}{\partial x}\right)_{yz} \delta x + \left(\frac{\partial \Phi}{\partial y}\right)_{xz} \delta y + \left(\frac{\partial \Phi}{\partial z}\right)_{xy} \delta z \tag{10.7}$$

and so on.

Incremental formulae, like eqns (10.6) and (10.7) are very useful because they lead naturally to many different connections between differential coefficients. If we divide eqn (10.6) by δu, for example, where the quantity u could be anything $(x, y, z,$ or something else), then we have

$$\frac{\delta z}{\delta u} \approx \left(\frac{\partial z}{\partial x}\right)_y \frac{\delta x}{\delta u} + \left(\frac{\partial z}{\partial y}\right)_x \frac{\delta y}{\delta u}$$

Taking the limit of $\delta u \to 0$ yields

$$\frac{dz}{du} = \left(\frac{\partial z}{\partial x}\right)_y \frac{dx}{du} + \left(\frac{\partial z}{\partial y}\right)_x \frac{dy}{du} \tag{10.8}$$

whereas the same procedure carried out at constant v gives

$$\left(\frac{\partial z}{\partial u}\right)_v = \left(\frac{\partial z}{\partial x}\right)_y \left(\frac{\partial x}{\partial u}\right)_v + \left(\frac{\partial z}{\partial y}\right)_x \left(\frac{\partial y}{\partial u}\right)_v \tag{10.9}$$

Perhaps the simplest illustration of eqn (10.9) is when $u = z$ and $v = y$ for then, after a little algebraic rearrangement, we find that

$$\left(\frac{\partial z}{\partial x}\right)_y = \frac{1}{(\partial x/\partial z)_y} \tag{10.10}$$

because $(\partial z/\partial z)_y = 1$ and $(\partial y/\partial z)_y = 0$; thus the reciprocity relationship for partial derivatives only holds as long as the same parameters are held fixed. Similarly, letting $u = x$ and $v = z$ leads to

$$\left(\frac{\partial z}{\partial y}\right)_x \left(\frac{\partial y}{\partial x}\right)_z \left(\frac{\partial x}{\partial z}\right)_y = -1 \tag{10.11}$$

where we have also made use of eqn (10.10).

Although both eqns (10.10) and (10.11) followed from eqn (10.9), we must not forget eqn (10.8) either. This is important because it links the so-called *total derivatives*, with ordinary d's, to the partial ones (having curly d's). Thus if $z = f(x, y)$, but x and y are themselves functions of t (time say), then eqn (10.8) can be used to evaluate dz/dt (by putting $u = t$) when direct substitution to obtain $z = f(t)$ is difficult.

The chain rule of eqn (10.9) is also useful for changing variables in expressions that involve partial differential coefficients. In the polar to Cartesian transformation, $x = r \cos\theta$ and $y = r \sin\theta$, for example, putting $u = r$ and $v = \theta$ in eqn (10.9) yields the relationship

$$\left(\frac{\partial z}{\partial r}\right)_\theta = \cos\theta \left(\frac{\partial z}{\partial x}\right)_y + \sin\theta \left(\frac{\partial z}{\partial y}\right)_x \qquad (10.12)$$

The substitution of $u = \theta$ and $v = r$ gives the complementary derivative $(\partial z/\partial\theta)_r$ in terms of $(\partial z/\partial x)_y$ and $(\partial z/\partial y)_x$. We need to be careful when evaluating higher-order derivatives, and remember how they are defined

$$\frac{\partial^2 z}{\partial r^2} = \frac{\partial}{\partial r_\theta}\left(\frac{\partial z}{\partial r}\right)_\theta = \cos\theta \frac{\partial}{\partial r_\theta}\left(\frac{\partial z}{\partial x}\right)_y + \sin\theta \frac{\partial}{\partial r_\theta}\left(\frac{\partial z}{\partial y}\right)_x$$

$$\left(\frac{\partial z}{\partial\theta}\right)_r = -r\sin\theta \left(\frac{\partial z}{\partial x}\right)_y + r\cos\theta \left(\frac{\partial z}{\partial y}\right)_x$$

where we have implicitly used the product rule of differentiation twice on the right-hand side. The awkward contributions, $\partial/\partial r_\theta(\partial z/\partial x)_y$ and $\partial/\partial r_\theta(\partial z/\partial y)_x$, can be ascertained by recognising, from eqn (10.12), that the differential operator $\partial/\partial r_\theta$ can be written as

$$\frac{\partial}{\partial r_\theta} = \cos\theta \frac{\partial}{\partial x_y} + \sin\theta \frac{\partial}{\partial y_x}$$

$$\frac{\partial}{\partial r_\theta}\left(\frac{\partial z}{\partial x}\right)_y = \cos\theta \frac{\partial^2 z}{\partial x^2} + \sin\theta \frac{\partial^2 z}{\partial x\partial y}$$

and knowing that it obeys all the usual rules of algebra, and applies to quantities on its immediate right.

Finally, we should note that an alternative to the use of the sort of chain rules discussed above is often provided by direct implicit differentiation. For example, the easiest way to obtain the partial derivative $(\partial y/\partial x)_z$ from $z^2 = x^3 y + \ln(y)$ is to apply the operator $\partial/\partial x_z$ to both sides of the equation

$$2z\left(\frac{\partial z}{\partial x}\right)_z = x^3\left(\frac{\partial y}{\partial x}\right)_z + 3x^2 y\left(\frac{\partial x}{\partial x}\right)_z + \frac{1}{y}\left(\frac{\partial y}{\partial x}\right)_z$$

Since $(\partial z/\partial x)_z = 0$ and $(\partial x/\partial x)_z = 1$, a little algebraic rearrangement shows that $(\partial y/\partial x)_z = -3x^2 y^2/(x^3 y + 1)$; this also follows from $(\partial z/\partial x)_y$, $(\partial z/\partial y)_x$ and eqns (10.10) and (10.11).

10.4 Taylor series

In chapter 6, we considered the Taylor series as a means of approximating the curve $y = f(x)$ locally by a low-order polynomial; this idea can be extended to functions of two or more variables. In fact, eqn (10.6) represents the first-order expansion for $z = f(x, y)$

$$f(x, y) = f(x_0, y_0) + (x - x_0)\left.\frac{\partial f}{\partial x}\right|_{x_0, y_0} + (y - y_0)\left.\frac{\partial f}{\partial y}\right|_{x_0, y_0} + \cdots$$

where the partial derivatives are evaluated at (x_0, y_0). This corresponds to approximating a two-dimensional surface by a tilted flat plane about the point of interest. A better estimate of the function is given by the further addition of the three terms involving the second derivatives of $f(x, y)$

$$\frac{1}{2}\left[(x-x_0)^2 \left.\frac{\partial^2 f}{\partial x^2}\right|_{x_0, y_0} + 2(x-x_0)(y-y_0) \left.\frac{\partial^2 f}{\partial x \partial y}\right|_{x_0, y_0} + (y-y_0)^2 \left.\frac{\partial^2 f}{\partial y^2}\right|_{x_0, y_0}\right]$$

which introduces a paraboloid contribution to the expansion. Higher-order terms become increasingly complicated, and are rarely used.

For functions of more than two variables, it's best to generalise the Taylor series by adopting a 'matrix-vector' notation

$$f(\underset{\sim}{x}) = f(\underset{\sim}{x}_0) + (\underset{\sim}{x} - \underset{\sim}{x}_0)^T \underset{\sim}{\nabla} f(\underset{\sim}{x}_0) + \frac{1}{2}(\underset{\sim}{x} - \underset{\sim}{x}_0)^T \underset{\sim}{\nabla}\underset{\sim}{\nabla} f(\underset{\sim}{x}_0)(\underset{\sim}{x} - \underset{\sim}{x}_0) + \cdots$$

where the column matrix $\underset{\sim}{x}$ has components $(x_1, x_2, x_3, \ldots, x_N)$, $\underset{\sim}{\nabla} f(\underset{\sim}{x}_0)$ is the N-dimensional gradient vector evaluated at the point $\underset{\sim}{x}_0$, and $\underset{\sim}{\nabla}\underset{\sim}{\nabla} f(\underset{\sim}{x}_0)$ is an $N \times N$ matrix whose ij^{th} element is given by the second partial derivative $\partial^2 f/\partial x_i \partial x_j$ (also calculated at $\underset{\sim}{x}_0$).

10.5   Maxima and minima

In section 4.6, we discussed the topic of maxima and minima for the curve $y = f(x)$; let's extend the analysis to encompass functions of several variables. The central idea of eqn (4.14), that a stationary point is a place where there is no slope, generalises very easily

$$\underset{\sim}{\nabla} f = 0 \tag{10.13}$$

the difference being that we now have to deal with a gradient vector rather than a single derivative. The only way in which a vector can be equal to zero is if all its components are nought. This leads to a set of N simultaneous equations: $\partial f/\partial x_i = 0$ for $i = 1, 2, \ldots, N$. Explicitly for the case of a two-parameter problem, $z = f(x, y)$, this means that

$$\left(\frac{\partial z}{\partial x}\right)_y = 0 \quad \text{and} \quad \left(\frac{\partial z}{\partial y}\right)_x = 0 \tag{10.14}$$

The best way to ensure that we find all the solutions, (x, y), of eqn (10.14) is to factorize the partial derivatives as much as possible.

As a concrete example of using eqn (10.14), consider the function $z = x^2 y^2 - x^2 - y^2$; the two conditions for its stationary points are

$$\left(\frac{\partial z}{\partial x}\right)_y = 2x(y^2 - 1) = 0 \quad \text{and} \quad \left(\frac{\partial z}{\partial y}\right)_x = 2y(x^2 - 1) = 0$$

The partial derivative $(\partial z/\partial x)_y$ is zero if either $x = 0$ or $y = \pm 1$; $(\partial z/\partial y)_x$ is nought if either $y = 0$ or $x = \pm 1$. The simultaneous equations are satisfied at five points, therefore: $(0, 0)$, $(1, 1)$, $(1, -1)$, $(-1, 1)$ and $(-1, -1)$.

Having found the stationary points with eqn (10.13), the next thing to do is to characterise their nature. In section 4.6, we did this by examining the sign of d^2y/dx^2. The situation is more complicated for functions of several variables because there are a large number of second partial derivatives. In fact

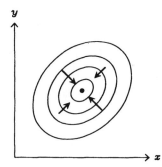

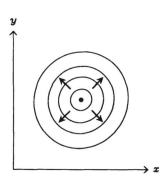

our discussion of the Taylor series expansion, in section 10.4, suggests that we need to look at the properties of the $\underset{\sim}{\nabla}\underset{\sim}{\nabla}f$ matrix. The paraboloid contribution introduced by the quadratic term yields a maximum if all the eigenvalues of $\underset{\sim}{\nabla}\underset{\sim}{\nabla}f$ (evaluated at the stationary point) are negative, and a minimum if they are entirely positive; eigenvalues of mixed signs indicate a *saddle point*, with the function increasing in some directions and decreasing in others.

$$\underset{\sim}{\nabla}\underset{\sim}{\nabla}z = \begin{pmatrix} \frac{\partial^2 z}{\partial x^2} & \frac{\partial^2 z}{\partial x\,\partial y} \\ \frac{\partial^2 z}{\partial x\,\partial y} & \frac{\partial^2 z}{\partial y^2} \end{pmatrix}$$

For our two-parameter case, $z = f(x,y)$, the eigenvalues of the 2×2 $\underset{\sim}{\nabla}\underset{\sim}{\nabla}z$ matrix do not need to be calculated explicitly. This is because we can make use of the fact that the product of the eigenvalues, $\lambda_1 \lambda_2$, is equal to the determinant of the real and symmetric $\underset{\sim}{\nabla}\underset{\sim}{\nabla}z$ matrix

$$\det\left(\underset{\sim}{\nabla}\underset{\sim}{\nabla}z\right) = \left(\frac{\partial^2 z}{\partial x^2}\right)\left(\frac{\partial^2 z}{\partial y^2}\right) - \left(\frac{\partial^2 z}{\partial x\,\partial y}\right)^2 \qquad (10.15)$$

Thus if $\det(\underset{\sim}{\nabla}\underset{\sim}{\nabla}z) > 0$, both eigenvalues must have the same sign and the stationary point is either a maximum or a minimum; a negative value indicates a saddle point, and a null result renders the test inconclusive (like $d^2y/dx^2 = 0$ in section 4.6). Having ascertained that $\det(\underset{\sim}{\nabla}\underset{\sim}{\nabla}z) > 0$, we need simply to look at the sign of $\partial^2 z/\partial x^2$ or $\partial^2 z/\partial y^2$, or their sum, to distinguish between a maximum and a minimum; it's negative for the former and positive for the latter.

Finally, carrying out the above analysis for our earlier example, where $\partial^2 z/\partial x^2 = 2(y^2 - 1)$, $\partial^2 z/\partial y^2 = 2(x^2 - 1)$, and $\partial^2 z/\partial x\,\partial y = 4xy$, we find that $(0,0)$ is a maximum, and the four points $(\pm1, \pm1)$ are saddle points.

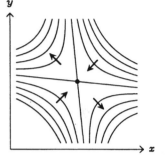

10.6 Constrained optimisation

Often we are not interested in simply finding the maximum or minimum of a function, but are obliged to do so subject to certain constraints. A scenario from everyday life that illustrates this point is the task of trying to buy the best computer, car, or whatever, given a limited budget. In thermodynamics, we could be trying to ascertain the populations of energy levels subject to a fixed number of particles, and the total energy of the system.

To explain the analysis further, suppose that we wish to find the 'optimum' value of the two-parameter function $z = f(x,y)$ subject to the condition that $g(x,y) = 0$. If we can rearrange the the constraint so that x can be written explicitly in terms of y, or vice versa, then it can be taken into account by an appropriate substitution in $f(x,y)$ to reduce it to a function of just one variable. For example, the stationary points of $z = x^2 - x + 2y^2$ subject to the condition that $x^2 + y^2 - 1 = 0$ are equivalent to the solutions of $dz/dx = 0$ where $z = x^2 - x + 2(1 - x^2)$ on putting $y^2 = 1 - x^2$ from the constraint; it leads to conditional maxima at $x = -1/2$ and $y = \pm\sqrt{3}/2$.

The procedure outlined above is of limited use because it is often very difficult to express x in terms of y, or the other way round, given $g(x,y) = 0$; there is also a further shortcoming in that some solutions can be missed. A more general approach for dealing with constrained optimisation is provided by the method of *Lagrange multipliers*. This states that our problem is equivalent to finding the ordinary (unconditional) stationary points of a new function, $F(x,y)$, constructed from $f(x,y)$ and $g(x,y)$ according to

$$F(x,y) = f(x,y) + \lambda\, g(x,y) \qquad (10.16)$$

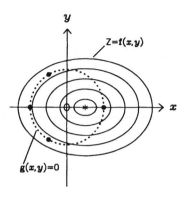

where λ is a constant, the Lagrange multiplier, whose value is as yet unknown. While the rational for eqn (10.16) is not easy to explain, its application is straightforward. The relevant values of x, y and λ can be obtained from the three simultaneous equations $(\partial F/\partial x)_y = 0$, $(\partial F/\partial y)_x = 0$, and $g(x,y) = 0$. For our earlier example, these are $2x(1+\lambda) - 1 = 0$, $2y(2+\lambda) = 0$, and $x^2 + y^2 - 1 = 0$ respectively. After some thought, and effort, the solutions are seen to be $x = -1/2$ and $y = \pm\sqrt{3}/2$, as before, with $\lambda = -2$, but also $x = \pm 1$ and $y = 0$, which are conditional minima with $\lambda = 1 \mp 1/2$.

Although we have illustrated the use of Lagrange multipliers for the simplest type of setup, the method generalises quite naturally to cases where there are several constraints or many variables in the function. Suppose we want to find the stationary points of $f(\underset{\sim}{x})$, where the vector $\underset{\sim}{x}$ has components $(x_1, x_2, \ldots, x_N)$, subject to $g_1(\underset{\sim}{x}) = 0$, $g_2(\underset{\sim}{x}) = 0$, $\ldots$, $g_M(\underset{\sim}{x}) = 0$; then, we begin by constructing the hybrid function $F(\underset{\sim}{x})$

$$F(\underset{\sim}{x}) = f(\underset{\sim}{x}) + \lambda_1 g_1(\underset{\sim}{x}) + \lambda_2 g_2(\underset{\sim}{x}) + \cdots + \lambda_M g_M(\underset{\sim}{x})$$

where the λ_i, for $i = 1$ to M, are Lagrange multipliers. The desired values are given by the $N+M$ parameters $x_1, x_2, \ldots, x_N, \lambda_1, \lambda_2, \ldots, \lambda_M$ which satisfy the $N+M$ set of simultaneous equations $\partial F/\partial x_1 = 0$, $\partial F/\partial x_2 = 0$, $\ldots$, $\partial F/\partial x_N = 0$, $g_1(\underset{\sim}{x}) = 0$, $g_2(\underset{\sim}{x}) = 0$, $\ldots$, $g_M(\underset{\sim}{x}) = 0$.

Exercises

10.1 Determine from first principles the partial derivatives $(\partial z/\partial x)_y$ and $(\partial z/\partial y)_x$ where $z(x,y) = x^3/(1-y)$. Evaluate (by any means) $\partial^2 z/\partial x^2$ and $\partial^2 z/\partial y^2$, and check that $\partial^2 z/\partial x\,\partial y = \partial^2 z/\partial y\,\partial x$.

10.2 If $f(x,y,z) = \cos(xyz)$, evaluate $\partial^3 f/\partial x\,\partial y\,\partial z$ in which the appropriate variables are held constant.

10.3 Verify that $x^2 = y^2 \sin(yz)$ satisfies $(\partial x/\partial y)_z (\partial y/\partial z)_x (\partial z/\partial x)_y = -1$.

10.4 Consider the function $f(x,y) = xy(1-y+x)$. Calculate the gradient vector, $\underline{\nabla} f$, at the points $(-1/2, 0)$, $(-1/2, 1/2)$ and $(0, 1/2)$. Find the stationary point which lies within the triangle bounded by the points $(-1, 0)$, $(0, 0)$ and $(0, 1)$. Sketch the function within the aforementioned triangle, marking in the directions of $\underline{\nabla} f$ where it has been calculated.

10.5 Show that if $f(u, v) = 0$, where $u = x + y$ and $v = x^2 + xy + z^2$, then $x + y = 2z[(\partial z/\partial y)_x - (\partial z/\partial x)_y]$.

10.6 Use the substitution $u = x + ct$ and $v = x - ct$ to reduce the wave equation $c^2 \partial^2 z/\partial x^2 = \partial^2 z/\partial t^2$ to the form $\partial^2 z/\partial u\,\partial v = 0$.

10.7 Find and classify all the (real) stationary values of the function $f(x,y) = y^2(a^2 + x^2) - x^2(2a^2 - x^2)$, where a is a constant.

10.8 Using the method of Lagrange multipliers, find the stationary values of the function e^{-xy} subject to the condition $x^2 + y^2 = 1$.

11 Line integrals

11.1 Line integrals

In this chapter, we will be considering the topic of *line integrals*. As a simple example of what they are and how they arise, we can think about an elementary formula which should be familiar from school physics: 'work done $=$ force $\times$ distance'. As we learnt in chapter 8, both force and distance moved are vectors, and so the times symbol should really be a scalar, or dot, product. Now, if the force is not constant, or the motion is not in a straight line, the situation will have to be analysed in terms of the sum of many small basic contributions

$$\text{Work done} \; = \int_{\text{path}} \underset{\sim}{F} \cdot \underset{\sim}{dl} \qquad (11.1)$$

where $\underset{\sim}{F}$ is the vectorial force field, $\underset{\sim}{dl}$ is a very short element of a straight line, and the whole movement takes place along a specified path.

If the motion is in a two-dimensional plane, $\underset{\sim}{F} = (F_x, F_y)$ and $\underset{\sim}{dl} = (dx, dy)$. Then, we have

$$\text{Work done} \; = \int_{\text{path}} F_x \, dx + F_y \, dy \qquad (11.2)$$

which can be written as the sum of two separate integrals (for dx and dy), where the components of the force, F_x and F_y, may be functions of x and y. The path will typically go from some point A, with coordinates (x_A, y_A), to B, at (x_B, y_B), along a trajectory defined by the curve $y = f(x)$. The integral of eqn (11.2) can be evaluated by expressing everything in it entirely in terms of x

$$\int_{\text{path}} F_x(x,y) \, dx + F_y(x,y) \, dy = \int_{x_A}^{x_B} \left[F_x \big(x, f(x) \big) + F_y \big(x, f(x) \big) f'(x) \right] dx$$

where we have used the fact that that the derivative $f'(x) = dy/dx$ to replace dy by $f'(x) \, dx$. We could, of course, have converted everything into y instead, if that was more convenient.

As a concrete illustration of the discussion above, suppose that we had to calculate the integral of $y^3 \, dx + x \, dy$ from $(0,0)$ to $(1,1)$ along two different paths: (a) $y = x^2$; and (b) the straight lines from $(0,0)$ to $(1,0)$, and from $(1,0)$ to $(1,1)$. For the first case, the substitution of $y = x^2$ and $dy = 2x \, dx$ leads to

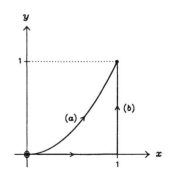

$$\int_{\text{path(a)}} y^3 \, dx + x \, dy = \int_0^1 \left(x^6 + 2x^2 \right) dx = \left[\frac{x^7}{7} + \frac{2x^3}{3} \right]_0^1 = \frac{17}{21}$$

For the two-part route we have $y = 0$ and $dy = 0$ between $(0,0)$ and $(1,0)$, and $x = 1$ and $dx = 0$ on going from $(1,0)$ to $(1,1)$; hence, the sum of these straight line contributions gives

$$\int_{\text{path(b)}} y^3 \, dx + x \, dy = \int_0^1 dy = [y]_0^1 = 1 \neq \frac{17}{21}$$

Thus the result of a line integral is generally dependent on the path followed from start to finish. The exception to this is when the integrand is an exact differential, which we will come to shortly, and then, as seen in section 11.2, the line integral is independent of the route taken, and a complicated path can be replaced by a simpler one to facilitate the calculation.

We should mention an alternative form of the line integral which is appropriate when dealing with scalar quantities

$$I = \int_{\text{path}} F(x,y) \, dl \tag{11.3}$$

where the integration is with respect to the arc-length, dl, along the curve $y = f(x)$; the path may also be specified by a parametric equation such as $x = g(t)$ and $y = h(t)$. The way to handle eqn (11.3) is very similar to eqn (11.2), except that dl is related to dx or dt by

$$dl = \sqrt{1 + \left(\frac{dy}{dx}\right)^2} \, dx \quad \text{or} \quad dl = \sqrt{\left(\frac{dx}{dt}\right)^2 + \left(\frac{dy}{dt}\right)^2} \, dt \tag{11.4}$$

which essentially follows from Pythagoras' theorem, $dl^2 = dx^2 + dy^2$; if $y = x^2$, for example, $dl = \sqrt{1 + 4x^2} \, dx$. It almost goes without saying that that $F(x,y)$ can be written as $F(x, f(x))$ or $F(g(t), h(t))$, depending on the format used for the integration path.

11.2 Exact differentials

Sometimes we are faced with expressions of the type $P(x,y) \, dx + Q(x,y) \, dy$, where P and Q are arbitrary functions of x and y, and a question that arises is whether this constitutes the formula for the increment df of some quantity $f(x,y)$. If it does, then $P \, dx + Q \, dy$ is said to be an *exact differential*; otherwise, it is not. So, how do we test for it?

If such an $f(x,y)$ existed, then a comparison with the limiting form of eqn (10.6) shows that

$$P(x,y) = \left(\frac{\partial f}{\partial x}\right)_y \quad \text{and} \quad Q(x,y) = \left(\frac{\partial f}{\partial y}\right)_x \tag{11.5}$$

$$df = \left(\frac{\partial f}{\partial x}\right)_y dx + \left(\frac{\partial f}{\partial y}\right)_x dy$$

While this must be true, it's not very helpful as it stands because it requires us to know $f(x,y)$ explicitly. A test for exactness which is independent of $f(x,y)$ does follow from eqn (11.5), however, when it is combined with eqn (10.5)

$$\left(\frac{\partial P}{\partial y}\right)_x = \left(\frac{\partial Q}{\partial x}\right)_y \tag{11.6}$$

Thus, for example, $3xy^2\,dx + 3x^2\,y\,dy$ is an exact differential but $\cos y\,dx + \sin x\,dy$ is not.

But why should we be interested in exact differentials at all? Well, to understand that, we need to consider the integral of $P\,dx + Q\,dy$ from some initial point A, with coordinates (x_A, y_A), to B, defined by (x_B, y_B). If $P\,dx + Q\,dy = df$, then

$$\int_A^B P\,dx + Q\,dy = \int_A^B df = \left[f(x,y)\right]_A^B = f(x_B, y_B) - f(x_A, y_A)$$

In other words, the integral depends only on the end-points, A and B, and not on the path taken for going from one to the other. If $P\,dx + Q\,dy$ is not an exact differential, the first step above does not follow; consequently, we must specify the details of the integration-path used and deal with any ensuing complications in the calculation.

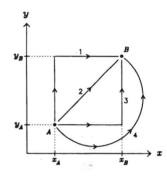

The physical relevance of the discussion in this section is that exact differentials are related to *conservative fields* (or forces) and *state functions*. For example, the change in the gravitational potential energy of an object depends only on the difference in its height, and not on any other characteristics of its motion. Similarly, in thermodynamics, the value of a function-of-state depends literally on the state (temperature, pressure and volume etc) of the system and not on how it got there.

Exercises

11.1 Show that the integral of $y^3\,dx + 3xy^2\,dy$ is independent of path by evaluating it over the two paths, (a) and (b) used in the example of section 11.1.

11.2 Evaluate the integral of $xy\,dl$ along the two paths, (a) and (b), used in the example of section 11.1.

11.3 Prove that if C_V is independent of volume V, $\delta q = C_V\,dT + (RT/V)\,dV$ is not an exact differential (where R is a constant). Show that by dividing this equation by T, it becomes exact. Comment on the relevance of this to thermodynamics.

12 Multiple integrals

12.1 Physical examples

In chapter 5, on integration, we were concerned with the area under the curve $y = f(x)$. Following our discussion about partial differentiation, however, we know that many functions of interest in real-life entail several variables. The topic of *multiple integrals* is, therefore, a natural extension of our earlier ideas to deal with multi-parameter problems.

To get a feel for how multiple integrals arise, let's consider a couple of physical examples. Suppose that we wish to calculate the force exerted on a wall by a gale. If the pressure P was constant across the whole face with area A, then the total force is simply $P \times A$. With a varying pressure $P(x,y)$, the answer is not so obvious. This situation can be handled by thinking about the wall as consisting of many small square segments, each with area $\delta x \, \delta y$, so that the total force is the sum of all the contributions $P(x,y) \, \delta x \, \delta y$; in the limiting case when $\delta x \to 0$ and $\delta y \to 0$, we have

$$\text{Force} \ = \ \iint\limits_{\text{wall}} P(x,y) \, \mathrm{d}x \, \mathrm{d}y \qquad (12.1)$$

where the *double integral* indicates that the infinitesimal summation is being carried out over a two-dimensional surface (in the x and y directions). Incidentally, if the wall does not have a conventional (rectangular) shape then its area can be similarly be calculated according to

$$\text{Area} \ = \ \iint\limits_{\text{wall}} \mathrm{d}x \, \mathrm{d}y \qquad (12.2)$$

The double integral is also called a *surface integral*.

Another illustration is provided by quantum mechanics where the modulus-squared of the wavefunction, $|\psi(x,y,z)|^2$, of an electron (say) gives the *probability density* of finding it at some point in space. The chances that the electron is in a small (cuboid) region of volume $\delta x \, \delta y \, \delta z$ is then $|\psi(x,y,z)|^2 \, \delta x \, \delta y \, \delta z$. Hence, the probability of finding it within a finite domain V is given by

$$\text{Probability} \ = \ \iiint\limits_{V} |\psi(x,y,z)|^2 \, \mathrm{d}x \, \mathrm{d}y \, \mathrm{d}z \qquad (12.3)$$

which is known as a *triple*, or *volume*, *integral*.

12.2 The order of integration

As a concrete example of how to calculate multiple integrals, let's consider a very easy case; namely, working out the formula for the area of a right-angled triangle. The simplest setup is to place one apex at the origin and have the base, of length L, along the x-axis; the vertical edge, of height H, can then rise up from $(L,0)$ to (L,H). If we take a set of small elementary boxes, of area $\delta x\,\delta y$, and stack them up parallel to the y-axis, at some point along the x-axis (where $0\le x\le L$), then we will obtain a narrow vertical strip going from $y=0$ to $y=Hx/L$; in the limit of $\delta x \to 0$ and $\delta y \to 0$, its area is given by

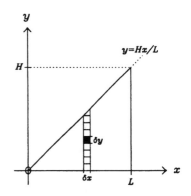

$$\text{Area of vertical strip} \;=\; \int_{y=0}^{y=\frac{Hx}{L}} dx\,dy \;=\; dx\int_{y=0}^{y=\frac{Hx}{L}} dy$$

where we have taken dx out of the y-integral because it does not depend on y. The area of the triangle is then the sum of all such contributions, with x ranging from $x=0$ to $x=L$

$$\text{Area of triangle} \;=\; \int_{x=0}^{x=L} dx \int_{y=0}^{\frac{Hx}{L}} dy \qquad (12.4)$$

where the convention is that the integral on the far right is evaluated first. Hence, the expected result emerges

$$\text{Area} \;=\; \int_0^L \big[y\big]_0^{Hx/L}\,dx \;=\; \frac{H}{L}\int_0^L x\,dx \;=\; \frac{H}{L}\left[\frac{x^2}{2}\right]_0^L \;=\; \frac{1}{2}HL$$

While this may seem like a rather tortuous way of getting to an obvious answer, the method would become essential if the shape was more complicated or the integral was of the type in eqn (12.1) instead of eqn (12.2); that is to say, with $P(x,y)$ not uniform. In a manner similar to partial derivatives, any occurrence of x inside inside the y-integral is treated like a constant (and vice versa).

$$\int_{y=0}^{y=\frac{\pi}{2}} \cos(xy)\,dy = \left[\frac{1}{x}\sin(xy)\right]_{y=0}^{y=\frac{\pi}{2}}$$
$$= \frac{1}{x}\sin\!\left(\frac{\pi x}{2}\right)$$

We could, of course, have carried out the calculation by stacking the elementary boxes along the x-direction first, going from $x=Ly/H$ to $x=L$, and then adding up the contributions from all the horizontal strips, so that the y-integral ranges from $y=0$ to $y=H$

$$\text{Area of triangle} \;=\; \int_{y=0}^{y=H} dy \int_{x=\frac{Ly}{H}}^{x=L} dx \qquad (12.5)$$

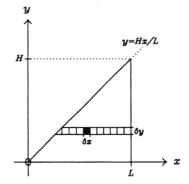

Upon evaluating the x-integral, followed by the one in y, we again obtain the expected result

$$\text{Area} \;=\; \int_0^H \big[x\big]_{Ly/H}^{L}\,dy \;=\; L\int_0^H\left(1-\frac{y}{H}\right)dy \;=\; L\left[y-\frac{y^2}{2H}\right]_0^H \;=\; \frac{1}{2}HL$$

This illustrates the fact that the order of integration in a multiple integral does not matter, and can be interchanged. The only thing we need to be careful about is the limits, because these usually need to be altered; for example, x varied between 0 and L in eqn (12.4) but goes from Ly/H to L in eqn (12.5). The best way of ascertaining the correct values for the limits is to sketch a diagram of the region being considered.

Although we noted that the order in which a multiple integral is carried out is up to us, the alternatives may entail different amounts of algebraic difficulty. As a general rule-of-thumb, it's normally advisable to do the easiest integral first and leave the hardest until the end.

12.3   Choice of coordinates

As a slightly more interesting variant on the triangle of the previous section, let's derive the formula for the area of a circle. By symmetry, we can reduce the problem to being one of four times the area of $x^2 + y^2 \leq R^2$ in the positive quadrant. Integrating first along y, and then adding together all the narrow vertical strips in the x-direction, we have

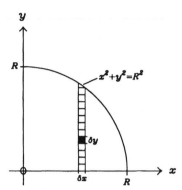

$$\text{Area of circle } = 4 \int_{x=0}^{x=R} dx \int_{y=0}^{y=\sqrt{R^2-x^2}} dy \qquad (12.6)$$

While the y-integral is straightforward, giving $\sqrt{R^2 - x^2}$, the second step is less so (requiring the substitution $x = R\sin\theta$). Reversing the order of integration does not help, but changing from Cartesian to polar coordinates does; rather than working in x and y, we can tackle the problem in r and θ (as in section 8.7).

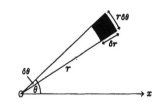

Remembering that we are trying to sum up small elements of area, the box generated when the polar coordinates are varied by a tiny amount has sides of length δr and $r\delta\theta$ (where the angle is in radians). In other words, we have

$$\text{Element of area } = dx\,dy = r\,dr\,d\theta \qquad (12.7)$$

With this relationship, we can then write eqn (12.6) as

$$\text{Area of circle } = 4 \int_{r=0}^{r=R} r\,dr \int_{\theta=0}^{\theta=\frac{\pi}{2}} d\theta \qquad (12.8)$$

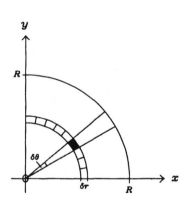

where the range $\theta = 0$ to $\pi/2$ and $r = 0$ to R covers the positive quadrant, irrespective of the order of integration. In eqn (12.8), the elementary boxes are first stacked up around the circle at a fixed radius; then these thin rings are added together, going from the centre to the circumference. The two integrals in eqn (12.8) are easily evaluated, and lead to the familiar result that the area of the circle is πR^2.

The example above highlights the point that multiple integrals, like many other mathematical operations, become much simpler if they are formulated in a coordinate system which matches the geometry of the problem. Although all our illustrations have been of double integrals, the generalisation to triple, and higher-order, ones is straightforward; in particular, the conclusions regarding the order of the integration and the choice of coordinates holds equally well. Thus Cartesian, cylindrical polar and spherical polar coordinates are best for problems having rectangular, cylindrical and spherical characteristics respectively. Apart from the limits of the integrals, the only thing we need to be careful about is the correct transformation of the elements of area, volume and so on. This is most easily done by drawing a diagram, as we did for two-dimensional polars, but can also be calculated by a *Jacobian* determinant

$$\mathrm{d}^M\mathrm{Vol} = \mathrm{d}x_1\,\mathrm{d}x_2\,\dots\,\mathrm{d}x_M = \begin{vmatrix} \frac{\partial x_1}{\partial X_1} & \frac{\partial x_1}{\partial X_2} & \cdots & \frac{\partial x_1}{\partial X_M} \\ \frac{\partial x_2}{\partial X_1} & \frac{\partial x_2}{\partial X_2} & \cdots & \frac{\partial x_2}{\partial X_M} \\ \vdots & \vdots & \ddots & \vdots \\ \frac{\partial x_M}{\partial X_1} & \frac{\partial x_M}{\partial X_2} & \cdots & \frac{\partial x_M}{\partial X_M} \end{vmatrix} \mathrm{d}X_1\,\mathrm{d}X_2\,\dots\,\mathrm{d}X_M \quad (12.9)$$

where M is the dimension of the 'volume' (so that $M = 2$ is an area), the lower-case x's are Cartesian components ($x_1 = x$, $x_2 = y$, $x_3 = z$, etc) and the capital X's are the alternative coordinates (e.g. $X_1 = r$, $X_2 = \theta$, and so on). In the two-dimensional Cartesian to polar transformation, for example, $x = r\cos\theta$ and $y = r\sin\theta$; hence eqn (12.9) gives a Jacobian of r, and we recover the result of eqn (12.7)

$$\begin{aligned} \mathrm{d}^3\mathrm{Vol} &= \mathrm{d}x\,\mathrm{d}y\,\mathrm{d}z & (\textit{Cartesian}) \\ &= r\,\mathrm{d}r\,\mathrm{d}\theta\,\mathrm{d}z & (\textit{Cylindrical}) \\ &= r^2\sin\theta\,\mathrm{d}r\,\mathrm{d}\theta\,\mathrm{d}\phi & (\textit{Spherical}) \end{aligned}$$

With the material in this section, we are now able to do the integral of a *Gaussian* function which was beyond us in chapter 5

$$I = \int_0^\infty e^{-x^2}\,\mathrm{d}x \quad (12.10)$$

The first step is non-intuitive, and consists of squaring I

$$I^2 = \int_{x=0}^{x=\infty}\int_{y=0}^{y=\infty} e^{-(x^2+y^2)}\,\mathrm{d}x\,\mathrm{d}y \quad (12.11)$$

where we have multiplied eqn (12.10) by a copy of itself, but with y as the dummy variable (instead of x). Changing from Cartesian to polar coordinates, according to eqn (12.7), eqn (12.11) becomes

$$I^2 = \int_{r=0}^{r=\infty} r\,e^{-r^2}\,\mathrm{d}r \int_{\theta=0}^{\theta=\frac{\pi}{2}}\,\mathrm{d}\theta$$

where the new limits still pertain to the positive quadrant. The product of the (now) two easy integrals yields $I^2 = \pi/4$; hence, on taking square roots, we find that eqn (12.10) is equal to $\sqrt{\pi}/2$.

$$\int_{-\infty}^{\infty} e^{-x^2/2\sigma^2}\,\mathrm{d}x = \sigma\sqrt{2\pi}$$

Exercises

12.1 Show that the area of the ellipse $(x/a)^2 + (y/b)^2 = 1$ is equal to $\pi a b$.

12.2 By drawing a (large) suitable diagram, show that an element of volume in spherical polar coordinates is given by $r^2\sin\theta\,\mathrm{d}r\,\mathrm{d}\theta\,\mathrm{d}\phi$; hence derive the formula for the volume of a sphere.

12.3 By using cylindrical polar coordinates, show that the volume of the solid generated when the curve $y = f(x)$, for $a \le x \le b$, is rotated about the x-axis through $360°$ is given by $\pi \int_a^b y^2\,\mathrm{d}x$.

12.4 Evaluate the double integral $\iint x^2(1 - x^2 - y^2)\,\mathrm{d}x\,\mathrm{d}y$ over a circle of radius 1 centred at $x = 0$ and $y = 0$ in two different coordinate systems: (i) Cartesian, and (ii) polar.

13 Ordinary differential equations

13.1 Definition of terms

An *ordinary* differential equation (ODE) is one that involves only normal differentials of $y(x)$, that is dy/dx, sometimes known as *the 'D' operator*, and its derivatives, and none of those nasty partial curly-d's which we discussed in chapter 10. The *order* of a differential equation is the highest derivative of y that appears in it: a first-order differential equation includes terms like $y' = dy/dx$ but no higher-order ones such as $y'' = d^2y/dx^2$ or $y''' = d^3y/dx^3$. We have already met the concept of *linearity* in chapter 4 which when applied to differential equations means that y and all its derivatives are raised to no power higher than the first; equations where this is not true are called *non-linear*. The *degree* of a differential equation is the power to which the highest derivative (of y) is raised, but we will only be dealing with those of the first.

13.2 First-order: separable

The most elementary ODE is

$$\frac{dy}{dx} = k \tag{13.1}$$

with k a constant, and is first-order, linear and of first-degree. As we learnt in chapter 4 this equation describes a function $y(x)$ which has a constant slope, so that its solution is simply a straight line. We can find the equation of this line by integrating both sides of eqn (13.1) with respect to x, giving

$$y = kx + A \tag{13.2}$$

Note that a single arbitrary constant A has appeared as a result of a single integration, and this introduces a result of crucial importance — the *general solution* of an nth order ODE must contain n arbitrary constants. It is easy to see why one, and only one, arbitrary constant is needed in this first-order case: eqn (13.1) tells us that $y(x)$ has a constant slope k, but contains nothing about the intercept of the associated line with $x = 0$. Hence any A will satisfy eqn (13.1), and we need a *boundary condition* to fix its value. For example, if we were told that $y = 0$ when $x = 0$ then we would know that $A = 0$, and would have the unique solution of a straight line of slope k passing through the origin.

A more general first-order ODE is

$$\frac{dy}{dx} = X(x)\,Y(y) \tag{13.3}$$

in which X and Y are general functions of x and y respectively. Just like eqn (13.1), this is a *separable* equation because, dividing both sides by Y and integrating with respect to x, we obtain

$$\int \frac{dy}{Y(y)} = \int X(x) \, dx \qquad (13.4)$$

which can be used to get a solution for $y(x)$. Let's illustrate how this works: if $X(x) = \cos x$ and $Y(y) = y$ then eqn (13.4) gives

$$\ln y = \sin x + A \quad \Longrightarrow \quad y = B e^{\sin x}$$

where an additive constant has been replaced by a multiplicative one, $B = e^A$, on taking exponentials.

13.3   First-order: homogeneous

A function $f(x, y)$ is *homogeneous* if it scales by a factor λ^m when both x and y are multiplied by λ (m is sometimes called the degree of the function, as opposed to that of an ODE). As an example $f = x^3 + 3x^2 y + 3y^2 x + 5y^3$ is a third-degree homogeneous function of x and y because if both x and y are doubled then f is increased by a factor 2^3. Any homogeneous function can be rewritten as $f = x^m F(V)$, where F is a function only of $V = y/x$: in our case, $F(V) = 1 + 3V + 3V^2 + 5V^3$. So next let's consider the ODE

$$\frac{dy}{dx} = \frac{\theta(x, y)}{\phi(x, y)} \qquad (13.5)$$

in which both $\theta(x, y)$ and $\phi(x, y)$ are homogeneous functions of the same degree. This means it can be expressed as

$$\frac{dy}{dx} = \frac{x^m \Theta(V)}{x^m \Phi(V)} = \Psi(V) \qquad (13.6)$$

where the factors of x^m have been cancelled, and, as before, we have $y = Vx$. Using the product rule of eqn (4.9), we can then differentiate $y = Vx$ with respect to x to obtain

$$\frac{dy}{dx} = V + x \frac{dV}{dx} = \psi(V) \qquad (13.7)$$

This can be re-arranged into the separable form of eqn (13.4)

$$\int \frac{dV}{\psi(V) - V} = \int \frac{dx}{x}$$

and hence solved for V. The final answer $y(x)$ is ascertained by using $y = Vx$, and a suitable boundary condition.

There are various 'change-of-variable tricks' which can transform some *inhomogeneous* first-order ODEs into a soluble homogeneous form. Consider the inhomogeneous first-order ODE

$$\frac{dy}{dx} = \frac{x + y + 1}{x - y}$$

By putting $u = x + a$ and $v = y + b$, and choosing the constants a and b to both be a half, which is the solution of the simultaneous equations $a + b = 1$ and $a = b$, the ODE above becomes

$$du = dx \quad \text{and} \quad dv = dy$$

$$\frac{dy}{dx} = \frac{dv}{du} = \frac{u+v}{u-v}$$

which is homogeneous, and can be solved in the way described earlier.

13.4   First-order: integrating factor

A general linear first-order ODE is of the type

$$\frac{dy}{dx} + yP(x) = Q(x) \tag{13.8}$$

and can be solved by multiplying both sides by an *integrating factor* $I(x)$

$$I(x) = \exp\left(\int P(x)\,dx\right) \tag{13.9}$$

Let's take as an example the equation

$$\frac{dy}{dx} + \frac{y}{x} = x$$

In this case $I(x) = \exp(\int dx/x) = \exp(\ln x) = x$, so that we get

$$x\left(\frac{dy}{dx} + \frac{y}{x}\right) = \frac{d}{dx}(yx) = x^2 \implies y = \frac{1}{x}\left(\frac{x^3}{3} + A\right)$$

$$y(x)\,I(x) = \int I(x)\,Q(x)\,dx$$

The critical step here is that with the integrating factor $I(x)$ calculated from eqn (13.9), we find that '$I(x)$ times the left-hand side of eqn (13.8) can be re-written as the differential of $I(x)$ times y'. It is also important to note that the arbitrary constant A is not simply tacked onto the end of the final answer.

Let us justify eqn (13.9) as the source of the integrating factor. Differentiating $I(x) \times y$, using the product rule of eqn (4.9), and equating it with $I(x)$ times the left-hand side of eqn (13.8), we have

$$\frac{d}{dx}(yI) = y\frac{dI}{dx} + I\frac{dy}{dx} = I\frac{dy}{dx} + IyP(x)$$

This simplifies to give a separable first-order ODE

$$\int \frac{dI}{I} = \int P(x)\,dx \tag{13.10}$$

which can easily be solved to yield eqn (13.9). Once again, changing variables can transform more complicated ODEs into the form of eqn (13.8).

13.5   Second-order: homogeneous

Consider next the second-order, linear homogeneous (right-hand side $= 0$) ODE

$$\frac{d^2y}{dx^2} + k_1\frac{dy}{dx} + k_2 y = 0 \tag{13.11}$$

where k_1 and k_2 are *constant coefficients*. We can guess that the form of the solution of this equation is $y = A\,e^{\alpha x}$ since, according to section 4.3, this is the only function that retains the same functional form on repeated differentiation. Substitution of this trial form into eqn (13.11) gives

$$A\,e^{\alpha x}\left(\alpha^2 + k_1\alpha + k_2\right) = 0 \implies \alpha^2 + k_1\alpha + k_2 = 0 \tag{13.12}$$

This quadratic is sometimes called the *auxiliary equation*, and solving it produces the allowed values of α. There are three cases of interest.

Case (i) is when there are two different real roots for α. For example

$$\frac{d^2y}{dx^2} - \omega_0^2 y = 0 \;\Rightarrow\; (\alpha + \omega_0)(\alpha - \omega_0) = 0 \;\Rightarrow\; y = Ae^{\omega_0 x} + Be^{-\omega_0 x} \quad (13.13)$$

where A and B are constants. We have used here the *principle of superposition*, a property of linear homogeneous ODEs that a general solution is obtained by taking a linear combination of all the allowed ones. Note that it is equally valid to express the general solution of eqn (13.13) as $y = C\cosh(\omega_0 x) + D\sinh(\omega_0 x)$ using the hyperbolic identities given in section 7.7, and defining new arbitrary constants $C = A + B$ and $D = A - B$.

As we have discussed in section 13.2, we need two arbitrary constants to obtain a general solution of second-order ODEs. Although it is intuitively obvious that A and B (or C and D) are the result of two integrations, we have obtained them here by the reverse approach of guessing the solution, and then differentiating it twice.

Case (ii) is when there are imaginary roots for α. For example

$$\frac{d^2y}{dx^2} + \omega_0^2 y = 0 \;\Rightarrow\; (\alpha + i\omega_0)(\alpha - i\omega_0) = 0 \;\Rightarrow\; y = Ae^{i\omega_0 x} + Be^{-i\omega_0 x}$$
$$(13.14)$$

This equation is of special physical importance as it describes the displacement (y) of many oscillating systems (where x is time); for example a pendulum. It is called the equation of *simple harmonic motion*. In this case the solution can also be written as $y = C\cos(\omega_0 x) + D\sin(\omega_0 x)$ using the trigonometric identities given in section 7.6, and defining $C = A + B$ and $D = i(A - B)$.

The introduction of an additional *damping term*, through a non-zero coefficient of dy/dx, complicates things only slightly; for example

$$\frac{d^2y}{dx^2} + \frac{dy}{dx} + y = 0 \;\Rightarrow\; y = \left[C\sin\left(\sqrt{3}x/2\right) + D\cos\left(\sqrt{3}x/2\right)\right]e^{-x/2} \quad (13.15)$$

$$\alpha = \frac{1}{2}\left(-1 + i\sqrt{3}\right)$$

The form of this solution shows why this system is described as being damped: the oscillation is multiplied by an exponential term which goes to zero as time (x) tends to infinity, meaning that the swing of the pendulum decays with time.

Finally, case (iii) is where the roots of the quadratic are equal, for example

$$\frac{d^2y}{dx^2} + 2\frac{dy}{dx} + y = 0 \;\Rightarrow\; (\alpha + 1)^2 = 0 \;\Rightarrow\; y = (A + B)e^{-x} = Ce^{-x} \quad (13.16)$$

This method of solution produces only one arbitrary constant ($C = A + B$), and a second functional form must be guessed. The correct form is $y = xe^{-\alpha x}$ which can be seen to give zero on substitution into eqn (13.16), giving the general solution $y = (A + Bx)e^{-x}$. This situation is sometimes called *critical damping*.

Some second-order, linear ODEs without constant coefficients can be transformed into the simple case of eqn (13.11) via a suitable change of variable. Consider, for example

$$x^2\frac{d^2y}{dx^2} + x\frac{dy}{dx} + y = 0 \quad (13.17)$$

Taking $u = \ln(x)$, and using the chain rule of eqn (4.12), we get

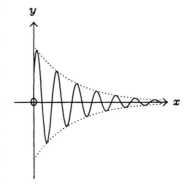

$$\frac{du}{dx} = \frac{1}{x}$$

$$x\frac{dy}{dx} = x\frac{dy}{du}\frac{du}{dx} = \frac{dy}{du} \tag{13.18}$$

$$x^2\frac{d^2y}{dx^2} = x^2\frac{d}{dx}\left(\frac{dy}{dx}\right) = x^2\frac{du}{dx}\frac{d}{du}\left(\frac{1}{x}\frac{dy}{du}\right) = \left(\frac{d^2y}{du^2} - \frac{dy}{du}\right) \tag{13.19}$$

Substitution of eqns (13.18) and (13.19) into eqn (13.17) transforms it into a constant-coefficient, second-order ODE in terms of y and u.

13.6   Second-order: inhomogeneous

The general second-order linear inhomogeneous ODE is

$$\frac{d^2y}{dx^2} + k_1\frac{dy}{dx} + k_2 y = F(x) \tag{13.20}$$

where $F(x)$ is any reasonable function of x. We can use the linearity of this ODE to find the solution by introducing a *complementary function* $c(x)$ and a *particular integral* $p(x)$, so that $y(x) = c(x) + p(x)$; $c(x)$ is the solution of the homogeneous function, that is eqn (13.20) with zero rather than $F(x)$ on the right-hand side . We have already learnt how to solve the ODE obeyed by $c(x)$ in section 13.5, because it is eqn (13.11) with $y(x) = c(x)$, and solving for it produces the two arbitrary constants required in our general solution for y. We can therefore concentrate on finding $p(x)$ which, when substituted into the left-hand side of eqn (13.20), must produce precisely $F(x)$ on the right-hand side with no further arbitrary constants.

There is an easy recipe for finding a suitable $p(x)$ for a given $F(x)$ although the algebra can sometimes be messy: construct $p(x)$ as a linear superposition of all the types of functions produced by $F(x)$ and its first and second derivatives. The values of the related coefficients are then determined by equating the pre-factors of each functional type on each side of eqn (13.20). An example should make this procedure clear

$$\frac{d^2y}{dx^2} + \omega_0^2 y = x\sin x \tag{13.21}$$

Following the method of section 13.5 we know that $c(x) = C\cos(\omega_0 x) + D\sin(\omega_0 x)$, and we need only to find $p(x)$. Our recipe suggests we try $p(x) = ax\sin x + bx\cos x + c\sin x + d\cos x$ since this is a linear combination of the only four types of functions which arise from the zero, first and second derivatives of $x\sin x$. Substituting into eqn (13.21), and equating coefficients of the four functional types, produces four simultaneous equations which can be solved to find values for the constants:

$$a(\omega_0^2 - 1) = 1 \quad (x\sin x)$$
$$b(\omega_0^2 - 1) = 0 \quad (x\cos x)$$
$$-2b + c(\omega_0^2 + 1) = 0 \quad (\sin x)$$
$$2a + d(\omega_0^2 - 1) = 0 \quad (\cos x)$$

in this case, $a = 1/(\omega_0^2 - 1)$, $b = 0$, $c = 0$, and $d = -2/(\omega_0^2 - 1)^2$.

Physically, eqn (13.20) represents an oscillating system driven by a force $F(x)$. In reality a damping term is almost always present so that the complementary function decays to zero as x, time, increases. This part of the solution, which is the only one that depends on the boundary conditions, is said to be *transient*; the particular integral gives the *steady state* solution. The steady state solution is often found using trial solutions which consist of a complex coefficient, or *complex amplitude*, multiplied by $\exp(imx)$.

We conclude this chapter by mentioning a difficulty sometimes encoun-tered with eqn (13.20). Consider the seemingly innocuous case: $k_1 = 0$, $k_2 = 1$, and $F(x) = \sin(x)$. The complementary function is the solution of eqn (13.14) with $\omega_0 = 1$, namely $y = C \sin x + D \cos x$. What do we try for the particular integral? Well, according to our recipe we try a linear combination of the set of functions found by differentiating $F(x)$. Unfortunately, this leads to a trial so-lution $p(x) = a \sin x + b \cos x$ which, since it is identical to the complementary function, produces zero when substituted into the left-hand side of eqn (13.20). How then do we produce non-vanishing $\sin x$ terms to match those in $F(x)$? Well, we need to guess functions that produce terms in $\sin x$ on differentiation. A reasonable first guess is $p(x) = ax \sin x + bx \cos x$. Substituting this trial so-lution into eqn (13.20), and equating coefficients, yields $a = 0$, $b = -1/2$, and thus a general solution $y = A \sin x + B \cos x - x \cos x/2$.

Exercises

13.1 The number N of radioactive atoms in a sample decays with time t ac-cording to the law $dN/dt = -\lambda N$. If there are originally N_0 radioactive atoms, obtain an expression for the *half-life*, the time required for the number of radioactive atoms to drop to $N_0/2$.

13.2 Obtain general solutions of the following first-order ODEs

(i) $\dfrac{dy}{dx} = \dfrac{1 - y^2}{x}$, (ii) $\dfrac{dy}{dx} = \dfrac{2y^2 + xy}{x^2}$, (iii) $\dfrac{dy}{dx} = \dfrac{x + y + 5}{x - y + 2}$,

(iv) $\dfrac{dy}{dx} + y \cot x = \operatorname{cosec} x$, (v) $\dfrac{dy}{dx} + 2xy = x$, (vi) $\dfrac{dy}{dx} + \dfrac{y}{x} = \cos x$.

13.3 The *Bernouilli* equation is

$$\frac{dy}{dx} + P(x)y = y^\alpha Q(x).$$

Transform it into the form of eqn (13.8) using the substitution $v = y^{-(\alpha-1)}$, and solve for $y(x)$ when $P(x) = x$, $Q(x) = x$, and $\alpha = 2$.

13.4 Show that eqn (13.8) can be written as $[Q(x) - P(x)y] \, dx - dy = 0$. By multiplying by $I(x)$, and applying the condition of eqn (11.6), show that $I(x)$ is given by eqn (13.9).

13.5 Obtain general solutions of eqn (13.20) when: (i) $k_1 = -2$, $k_2 = -3$, $F(x) = \sin x$; (ii) $k_1 = -2$, $k_2 = -8$, $F(x) = x^2$; (iii) $k_1 = 0$, $k_2 = \omega_0^2$, $F(x) = \cos(\omega x)$; (iv) $k_1 = 1$, $k_2 = 1$, $F(x) = \cos(\omega x)$; (v) $k_1 = 0$, $k_2 = 4$, $F(x) = \cos(2x)$. Obtain a full solution for case (i) given boundary con-ditions $y(0) = 0$, and y remains finite as $x \to \infty$. What happens in case (iii) when $\omega = \omega_0$? Find the steady state solution for case (iv) by express-ing the right-hand side as the real part of a complex exponential, and by trying $p = \mathcal{R}e\{A \exp(i\omega x)\}$.

13.6 Solve

$$x^2 \frac{d^2y}{dx^2} + 3x \frac{dy}{dx} + y = 0$$

using (i) the trial solution $y = Ax^\lambda$, and (ii) the method of section 13.5.

14 Partial differential equations

14.1 Elementary cases

The previous chapter, on ordinary differential equations, was restricted to functions of a single variable; now let's extend the analysis to deal with multi-parameter problems. As we learnt in chapter 10, such a generalisation entails the replacement of ordinary derivatives (like dy/dx) with partial ones (such as $\partial y/\partial x$). Hence, the material in this chapter goes under the heading of *partial differential equations*; or PDEs for short. For clarity, most of the examples will involve functions of just two variables.

Let's begin with the simplest PDE of all

$$\left(\frac{\partial z}{\partial x}\right)_y = 0 \tag{14.1}$$

If a derivative was zero, we would normally deduce that its integral was equal to a constant. In eqn (14.1), y is held fixed, and so any function of it, say $g(y)$, is equivalent to a constant in; hence, the most general solution is

$$z(x,y) = g(y)$$

This elementary example highlights that PDEs give rise to 'functions of integration', in contrast to ODEs which have 'constants of integration'.

Now let's consider a more interesting case. Suppose that we have the exact differential $df = [(x-y)/x^2]\,dx + [1/x]\,dy$; what is $f(x,y)$? Well, eqn (11.5) shows that

$$\left(\frac{\partial f}{\partial x}\right)_y = \frac{x-y}{x^2} \quad \text{and} \quad \left(\frac{\partial f}{\partial y}\right)_x = \frac{1}{x} \tag{14.2}$$

Starting with the second expression, because it's less complicated, it states that the derivative of $f(x,y)$ with respect to to y, with x treated like a constant, is equal to $1/x$; therefore, we can infer that

$$f(x,y) = \frac{y}{x} + g(x) \tag{14.3}$$

The best way to get a handle on $g(x)$ is to differentiate eqn (14.3) with respect to x, with y held fixed, and compare the result with the first part of eqn (14.2)

$$\left(\frac{\partial f}{\partial x}\right)_y = \frac{dg}{dx} - \frac{y}{x^2} = \frac{1}{x} - \frac{y}{x^2}$$

where we have replaced $(\partial g/\partial x)_y$ with dg/dx since, by definition, $g(x)$ is a function of x only. Thus $dg/dx = 1/x$, and so $g(x) = \ln x + C$ where C is a true constant. Hence, the solution to our exact differential is

$$f(x,y) = \frac{y}{x} + \ln x + C$$

which can, and should, be checked by making sure that it returns eqn (14.2).

For the last illustration in this section, let's consider an easy second-order example: $\partial^2 z/\partial x \partial y = 0$. This can be tackled by thinking about what the mixed partial derivative stands for

$$\frac{\partial^2 z}{\partial x \partial y} = \frac{\partial}{\partial x_y}\left(\frac{\partial z}{\partial y}\right)_x = 0 \qquad (14.4)$$

Integrating eqn (14.4) with respect to x, while treating y like a constant, yields

$$\left(\frac{\partial z}{\partial y}\right)_x = g(y)$$

A second integration, with respect to y at fixed x, gives the general solution as

$$z(x,y) = G(y) + h(x) \qquad (14.5)$$

where $G(y) = \int g(y)\,dy$ is also a function of y only. While eqn (14.5) may seem uninformative, since $G(y)$ and $h(x)$ cannot be determined without suitable boundary conditions, it does tell us that $z(x,y)$ is the sum of separate functions of x and y; that is to say, there can be no mixed terms.

All the PDEs in this section were elementary, in that they could be solved by direct integration as long as we thought carefully about what the partial derivatives actually meant. Now let's turn to a more realistic situation.

14.2 Separation of variables

Most PDEs of physical interest, such as the *wave equation*, tend to be of second order

$$\frac{\partial^2 y}{\partial x^2} = \frac{1}{c^2}\frac{\partial^2 y}{\partial t^2} \qquad (14.6)$$

where y could be the 'vertical' displacement of a string, x the position along it, t is time, and c the speed of the wave. There are several different ways of solving PDEs, but almost all of them are beyond the scope of this book. We will be focussing on just one particular method called the *separation of variables*; any similarity in the name with the approach described in section 13.2 is purely incidental, as they are quite unrelated.

Before coming to the main part of this section, however, we should mention an interesting point about eqn (14.6). The substitution of $u = x + ct$ and $v = x - ct$ transforms the wave equation into the form met in eqn (14.4), $\partial^2 y/\partial u \partial v = 0$; showing this was the task of exercise 10.6. Hence, the general solution of eqn (14.6) can be written as: $y(x,t) = G(x+ct) + h(x-ct)$, where the functions G and h must be determined from suitable boundary conditions.

The method of the separation of variables is best explained with reference to a specific example. The one we will use can be stated in terms of the following question: "A function $f(x,y)$ is defined in the strip $x \geq 0$ and $0 \leq y \leq a$, and obeys *Laplace's equation*

$$\frac{\partial^2 f}{\partial x^2} + \frac{\partial^2 f}{\partial y^2} = 0 \qquad (14.7)$$

If $f(x,y) \to 0$ as $x \to \infty$, and satisfies the boundary conditions

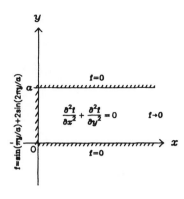

$$\left(\frac{\partial f}{\partial x}\right)_y = Y(y)\left(\frac{\partial X}{\partial x}\right)_y = Y\frac{dX}{dx}$$

$$\left(\frac{\partial f}{\partial y}\right)_x = X(x)\left(\frac{\partial Y}{\partial y}\right)_x = X\frac{dY}{dy}$$

$$f(x,0) = f(x,a) = 0$$
$$f(0,y) = \sin(\pi y/a) + 2\sin(2\pi y/a) \qquad (14.8)$$

find the appropriate solution of Laplace's equation."

As a useful preliminary to tackling this sort of problem, it is often helpful to sketch a diagram which highlights the region of interest, and shows the relevant boundary conditions. The first step in the separation of variables, and the thing that gives the method its name, is to try a solution of the form

$$f(x,y) = X(x)Y(y) \qquad (14.9)$$

We should emphasise that this simply represents an educated guess as to what the solution of eqn (14.7) might look like, a product of independent functions of x and y, X and Y, but we're not sure that it is correct; hence we preface eqn (14.9) with the word 'try' rather than 'let' or 'put'.

To see whether eqn (14.9) works, we need to substitute it into eqn (14.7) and check that a viable solution emerges

$$Y\frac{d^2X}{dx^2} + X\frac{d^2Y}{dy^2} = 0$$

where we have replaced the partial derivatives with ordinary ones because, by the definition of eqn (14.9), X and Y are respectively functions of x and y only. Dividing the expression above by XY, and rearranging the result, we obtain

$$\frac{1}{X}\frac{d^2X}{dx^2} = -\frac{1}{Y}\frac{d^2Y}{dy^2} \qquad (14.10)$$

At this stage, there is a fairly subtle point that requires some thought: the left and right-hand sides of eqn (14.10) are separately arbitrary functions of x and y, and yet they are always equal to each other. This can only happen if both are individually equal to a constant. Calling the latter ω^2, for reasons that will become obvious later, our original PDE of eqn (14.17) splits up into two equivalent ODEs

$$\frac{d^2X}{dx^2} = \omega^2 X \quad \text{and} \quad \frac{d^2Y}{dy^2} = -\omega^2 Y \qquad (14.11)$$

These can either be solved by using the material in section 13.5, that is by trying $X = A e^{px}$ and $Y = B e^{qy}$, and so on, or by recognising the relationship between eqn (14.11) and *simple harmonic motion*. In any case, eqns (14.9) and (14.11) combine to give the form of the solutions of eqn (14.7) as

$$f(x,y) = [A e^{\omega x} + B e^{-\omega x}][C\cos(\omega y) + D\sin(\omega y)] \qquad (14.12)$$

where A, B, C, and D are constants of integration.

Having dealt with the PDE *per se*, the remainder of our task consists of applying the boundary conditions. Starting with $f(x,y) = 0$ as $x \to \infty$, this can only be satisfied if $A = 0$ (otherwise f blows up); similarly, $f(x,0) = 0$ requires that $C = 0$ because the alternative of $B = 0$ would lead to the unacceptable conclusion that $f(x,y) = 0$. Next, $f(x,a) = 0$ means that $\sin(\omega a) = 0$; hence $\omega a = n\pi$, where n is an integer. The fact that n, and in turn ω, can take on a multitude of discrete values shows that there are a large number of solutions of the form of in eqn (14.12). Labelling each with the suffix n, the most general solution of eqn (14.7) now allowed is

$$f(x,y) = \sum_{n=-\infty}^{\infty} E_n \sin(n\pi y/a)\,e^{-n\pi x/a} \tag{14.13}$$

where we have consolidated the product $(BD)_n$ into a single coefficient, E_n, and implicitly set $A_n = C_n = 0$. The reason for the summation in eqn (14.13) is that the nature of the PDEs which we tend to encounter, such as eqn (14.7), have the property that a linear combination of solutions is also a solution. By comparing $f(0,y)$, term-by-term, with the expression in the boundary condition of eqn (14.8), we reach the final result with $E_1 = 1$, $E_2 = 2$, and all other $E_n = 0$

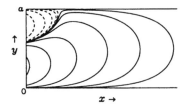

$$f(x,y) = \sin(\pi y/a)e^{-\pi x/a} + 2\sin(2\pi y/a)e^{-2\pi x/a} \tag{14.14}$$

Strictly speaking, we should now check that eqn (14.14) satisfies both the PDE of eqn (14.7) and the requirements of eqn (14.8).

Although we have illustrated the method of the separation of variables with a specific example, the steps for dealing with other problems are virtually identical. The power of the technique lies in its ability to reduce a PDE to an equivalent set of ODEs. On a practical note, the choice of the constant used at the stage of eqns (14.10) and (14.11), and the subsequent order in which the boundary conditions are implemented, plays an important part in facilitating a successful outcome. In this context, the recommended preliminary diagram is helpful because it suggests that the solution of the PDE will be a decaying function in the x direction and an oscillatory one in y; the use of a constant other than ω^2 would have led to more complex algebra, involving square root signs and quite possibly the imaginary i. It is also best to apply the simplest boundary conditions first, especially if they eliminate some of the terms at the stage of eqn (14.12), and leave the most complicated till the end.

14.3 Common physical examples

We have already met two PDEs of physical interest in section 14.2; namely, the wave equation of eqn (14.6) and Laplace's equation in eqn (14.7). While the former needs no introduction, the latter, and its more general form in the shape of *Poisson's equation* (where the right-hand side is not necessarily equal to zero), is of importance in *electrostatics*, *hydrodynamics*, and so on. A third common case is the *diffusion equation*

$$K\,\frac{\partial^2 u}{\partial x^2} = \frac{\partial u}{\partial t} \tag{14.15}$$

which describes the flow of heat in a metal, the spread of a solute in a solvent, and the like. The one-dimensional wave equation of eqn (14.6) can be extended to three dimensions by replacing the differential operator $\partial^2/\partial x^2$ with ∇^2

$$\nabla^2\psi = \frac{1}{c^2}\frac{\partial^2\psi}{\partial t^2} \tag{14.16}$$

where $\nabla^2\psi = \partial^2\psi/\partial x^2 + \partial^2\psi/\partial y^2 + \partial^2\psi/\partial z^2$; similarly, the diffusion equation of eqn (14.15) becomes $K\nabla^2 u = \partial u/\partial t$, and Laplace's equation of eqn (14.7) reduces to $\nabla^2 f = 0$. In solving eqn (14.16) with the separation of variables,

$$\nabla^2 \psi = \frac{\partial^2 \psi}{\partial x^2} + \frac{\partial^2 \psi}{\partial y^2} + \frac{\partial^2 \psi}{\partial z^2}$$

Cartesian

$$= \frac{1}{r^2} \frac{\partial}{\partial r}\left(r^2 \frac{\partial \psi}{\partial r}\right)$$
$$+$$
$$\frac{1}{r^2 \sin\theta} \frac{\partial}{\partial \theta}\left(\sin\theta \frac{\partial \psi}{\partial \theta}\right)$$
$$+$$
$$\frac{1}{r^2 \sin^2\theta} \frac{\partial^2 \psi}{\partial \phi^2}$$

Spherical polars

for example, we now try $\psi(x,y,z,t) = X(x)\,Y(y)\,Z(z)\,T(t)$. If the problem being tackled has an inherent spherical symmetry, such as the hydrogen atom, then, as noted in section 12.3, it is better to work in spherical polar coordinates rather than Cartesians. Writing $\nabla^2 \psi$ in terms of r, θ, and ϕ (which is much more complicated than its x, y, and z counterpart), we now try $\psi(r,\theta,\phi,t) = R(r)\,\Theta(\theta)\,\Phi(\phi)\,T(t)$. The radial part of the solution turns out to be related to *Bessel functions*, and the combined angular contribution is closely linked with *spherical harmonics*. With the application of suitable boundary conditions, a range of discrete possibilities emerges as in eqn (14.13); the suffices used to denote the alternatives have a direct correspondence with the various *quantum numbers* that appear in atomic physics.

Finally, as a matter of passing interest, we note that *Schrödinger's* (time-independent) *equation*, which is really a wave equation for 'standing waves', is also an eigenvalue equation

$$H\psi = E\psi \qquad (14.17)$$

where the *Hamiltonian* operator, H, is the sum of the kinetic energy, $-h^2\nabla^2/(8\pi^2 m)$, and the potential energy, V, and the constant E is the energy. The wavefunctions, ψ, which satisfy eqn (14.17) are called *eigenfunctions*, and represent the stationary states of the system, and the corresponding values of E, which are the eigenvalues, give the related energy levels; essentially, eqn (14.17) is the continuum form of the matrix definition of eqn (9.17).

Exercises

14.1 For the exact differential $df = y\cos(xy)\,dx + [x\cos(xy) + 2y]\,dy$, find $f(x,y)$.

14.2 Verify that $u(x,t) = \exp(-x^2/4kt)/\sqrt{4kt}$ is a solution of the diffusion equation $\partial u/\partial t = k\,\partial^2 u/\partial x^2$.

14.3 Given the Schrödinger equation for free electrons in two dimensions,

$$\frac{\partial^2 \psi}{\partial x^2} + \frac{\partial^2 \psi}{\partial y^2} + \frac{8\pi^2 mE\psi}{h^2} = 0$$

obtain the wavefunctions ψ, and allowed energy levels E, subject to the boundary conditions $\psi = 0$ at $x = 0$, $x = a$, $y = 0$, and $y = b$.

14.4 Laplace's equation in plane polar coordinates (r,θ) is

$$\frac{\partial^2 \Phi}{\partial r^2} + \frac{1}{r}\frac{\partial \Phi}{\partial r} + \frac{1}{r^2}\frac{\partial^2 \Phi}{\partial \theta^2} = 0$$

Show, by separating the variables, that there are solutions of the form

$$\Phi(r,\theta) = (A_0\theta + B_0)(C_0 \ln r + D_0) \qquad \text{and}$$

$$\Phi(r,\theta) = [A_p \cos(p\theta) + B_p \sin(p\theta)](C_p r^p + D_p r^{-p})$$

where the A's, B's, C's, D's and p are constants. If Φ is a single-valued function of θ, how does this restrict p? Solve the equation for $0 \leq r \leq a$ and (i) $\Phi(a,\theta) = T\cos\theta$, and (ii) $\Phi(a,\theta) = T\cos^3\theta$.

15 Fourier series and transforms

15.1 Approximating periodic functions

In chapter 6, we saw how a Taylor series could be used to mimic an arbitrary function by a simple low-order polynomial in the neighbourhood of a particular point. An alternative approximation is provided by a Fourier series, which is appropriate when the curve of interest is periodic

$$f(x) \approx a_0/2 + a_1 \cos(\omega x) + a_2 \cos(2\omega x) + a_3 \cos(3\omega x) + \cdots$$
$$+ b_1 \sin(\omega x) + b_2 \sin(2\omega x) + b_3 \sin(3\omega x) + \cdots \quad (15.1)$$

where the a's, b's and ω are constants; in fact, eqn (15.1) is the counterpart of eqn (6.1). That is to say, any repeating function can be represented as a sum of sines and cosines. We could, of course, replace $a_1 \cos(\omega x) + b_1 \sin(\omega x)$ with $c_1 \cos(\omega x + \phi_1)$ or $c_1 \sin(\omega x + \phi_1)$, but the former choice is usually preferable due to its linearity. The x-period of eqn (15.1) is given by the lowest *harmonic*, ω, as $2\pi/\omega$.

By multiplying eqn (15.1) separately by $\cos(n\omega x)$ and $\sin(n\omega x)$, and integrating both sides over one repeat interval, it can be shown that the Fourier coefficients are given by

$$a_n = \frac{\omega}{\pi} \int_0^{2\pi/\omega} f(x) \cos(n\omega x)\, dx \quad \text{and} \quad b_n = \frac{\omega}{\pi} \int_0^{2\pi/\omega} f(x) \sin(n\omega x)\, dx \quad (15.2)$$

where $n = 1, 2, 3, \ldots$, and relies on the fact that sines and cosines are orthogonal. In other words, the integral of $\sin(m\omega x)\sin(n\omega x)$ and $\cos(m\omega x)\cos(n\omega x)$ from $x = 0$ to $x = 2\pi/\omega$ is zero if $m \neq n$; and is always nought over this range for any mixed combination of sines and cosines. The reason for the odd-looking factor of a half with the a_0 in eqn (15.1) is that the formulae of eqn (15.2) then work even for $n = 0$; the corresponding term for b_0 has been omitted because its value turns out to be zero.

In practice, a Fourier series can be used to approximate a function even though it isn't actually periodic. The proviso is that we should only be interested in $f(x)$ over a limited range, $0 \leq x \leq L$ say. Then we can make it pseudo-repeating by arbitrarily defining $f(x) = f(x + jL)$, where j is an integer, or by letting $f(x) = f(-x)$ and $f(x) = f(x + 2jL)$, or by setting $f(x) = -f(-x)$ and $f(x) = f(x + 2jL)$, and so on; if the period is L then $\omega = 2\pi/L$, and if it's $2L$ then $\omega = \pi/L$. While each possibility will lead to a different Fourier expansion, so that there are no sine terms if $f(x) = f(-x)$ and no cosine factors if $f(x) = -f(-x)$, they will all give a reasonable approximation in the important region $0 \leq x \leq L$.

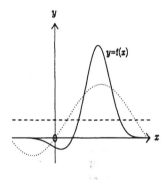

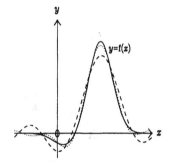

15.2   Taylor versus Fourier series

While a Taylor series provides a very good representation of $y = f(x)$ near the expansion point $x = x_0$, it soon yields a poor one as $|x - x_0|$ becomes large. By comparison, a Fourier series does not do an accurate job anywhere but mimics the overall shape of the curve quite well. The general characteristics of the latter, with its high-frequency oscillatory behaviour, means that a term-by-term differentiation is an extremely ill-advised operation; a corresponding integration is a much safer procedure, as the wiggles tend to cancel out. At a more technical level, a Taylor series requires that $f(x)$ be differentiable many times over where as the function, and its gradient $f'(x)$, need only be continuous for a Fourier series.

15.3   The Fourier integral

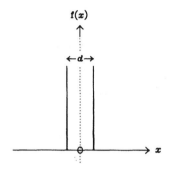

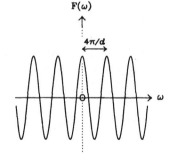

A more compact version of eqn (15.1) can be obtained by using the complex number relationship of eqn (7.11), which ties together sines and cosines through the imaginary exponential

$$f(x) = \sum_{n=-\infty}^{\infty} c_n\, e^{in\omega x} \tag{15.3}$$

where the coefficients, c_n, are complex. If necessary, the c's can be related to the a's and b's in eqn (15.1) with $a_n = (c_n + c_{-n})$ and $b_n = i(c_n - c_{-n})$.

The advantage of working in the complex notation is that it is much simpler to generalise when considering the continuum limit. That is to say, if the repeat interval of the Fourier series becomes infinitely long, so that $f(x)$ is no longer required to be periodic, then the summation of eqn (15.3) can be replaced with an integral

$$f(x) = \int_{-\infty}^{\infty} F(\omega)\, e^{i\omega x}\, d\omega \tag{15.4}$$

where the 'coefficient function', $F(\omega)$, is given by

$$F(\omega) = \frac{1}{2\pi} \int_{-\infty}^{\infty} f(x)\, e^{-i\omega x}\, dx \tag{15.5}$$

and can be regarded as the complex, and limiting, version of eqn (15.2). In eqns (15.4) and (15.5), $F(\omega)$ is known as the *Fourier transform* of $f(x)$; and $f(x)$ is the *inverse* Fourier transform of $F(\omega)$. Actually, which one we call the transform, and which one the inverse, is just a matter of convention; the important thing is that they come as a related pair, with opposite signs in the exponential. Even the detailed definition, in terms of the location of π, can vary: sometimes both eqns (15.4) and (15.5) are written with integral pre-factors of $1/\sqrt{2\pi}$, making them look more symmetrical, while it's also possible to have an $i2\pi\omega x$ in the exponent and nothing else.

The Fourier transform of $f(x)$, $F(\omega)$, is a 'spectrum' which can be plotted graphically as a function of ω. Its amplitude essentially tells us how much a sinusoidal curve of frequency ω contributes towards the synthesis of $f(x)$. By

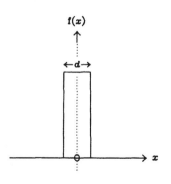

substituting $f(x) = \delta(x - d/2) + \delta(x + d/2)$ in eqn (15.5), for example, where $\delta(x - x_0)$ is a *delta-function* representing a very narrow spike of unit area at $x = x_0$, so that $\int \delta(x - x_0) \exp(-i\omega x) \, dx = \exp(-i\omega x_0)$, it is readily shown that the Fourier transform of two sharp peaks separated by a distance d is proportional to $\cos(\omega d/2)$. Similarly, putting $f(x) = 1$ for $-d/2 \le x \le d/2$ and zero otherwise leads to the result that the Fourier transform of a 'top-hat', or 'box', function is related to the *sinc function*, $\sin(\theta)/\theta$, where $\theta = \omega d/2$. What both these cases indicate is that the width of the features in $F(\omega)$ is inversely proportional to the characteristic spread of the structure in $f(x)$: the narrower the function the broader its Fourier transform, and vice versa.

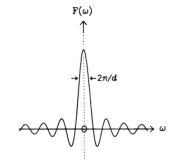

$F(\omega)$

$\leftarrow 2\pi/d$

ω

15.4 Some formal properties

Hitherto we have implicitly assumed that $f(x)$ is a real function, in that $f(x) = f(x)^*$, and can be plotted as a curve on a piece of paper, but there is nothing to stop it from being complex as far as eqns (15.4) and (15.5) are concerned. Indeed, $F(\omega)$ is generally complex. If $f(x)$ is real, as it usually is, then its Fourier transform can be shown to be *conjugate symmetric*; that is, $F(\omega) = F(-\omega)^*$. Other properties of $F(\omega)$ can also be derived when $f(x)$ exhibits symmetrical behaviour: $F(\omega) = F(-\omega)$ if $f(x)$ is an even function, so that $f(x) = f(-x)$; and $F(\omega) = -F(-\omega)$ when $f(x)$ is odd, or $f(x) = -f(-x)$. These features can be combined in that $F(\omega)$ is both real and symmetric if $f(x)$ is too, and is real and odd when $f(x)$ is as well.

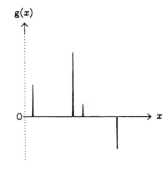

$g(x)$

0

x

One of the most important results in Fourier analysis is known as the *convolution theorem*. It states that

$$f(x) = g(x) \otimes h(x) \quad \Longleftrightarrow \quad F(\omega) = 2\pi \, G(\omega) \times H(\omega) \qquad (15.6)$$

where the $\otimes$ denotes a convolution, and $F(\omega)$, $G(\omega)$ and $H(\omega)$ are the Fourier transforms of $f(x)$, $g(x)$ and $h(x)$ respectively. A convolution between two functions is defined by the integral

$$g(x) \otimes h(x) = \int_{-\infty}^{\infty} g(y) \, h(x - y) \, dy = \int_{-\infty}^{\infty} g(x - y) \, h(y) \, dy \qquad (15.7)$$

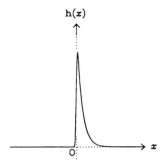

$h(x)$

0

x

and represents a blurring of $g(x)$ by $h(x)$, or vice versa. If $g(x)$ contains sharp, spiky, structure and $h(x)$ is a broad bell-shaped Gaussian function, for example, then $g(x) \otimes h(x)$ will be a smeared-out version of $g(x)$. The usefulness of eqn (15.6) lies in its ability to transform a potentially difficult integral in x-space into a straightforward product in ω, or Fourier, space.

The final property of Fourier transforms that we will consider concerns the *auto-correlation function*, or ACF, which provides information on the distance distribution of the various structures in $f(x)$. In other words, if there are two spikes separated by a distance L in $f(x)$, and with amplitudes A_1 and A_2, then they will contribute a symmetric pair of very sharp components at $x = \pm L$, and magnitude $A_1 A_2$, towards the ACF of $f(x)$. It can also be shown that the ACF of $f(x)$ is related to the Fourier transform of $|F(\omega)|^2 = F(\omega) \, F(\omega)^*$

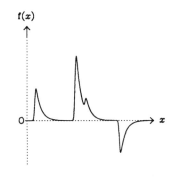

$f(x)$

0

x

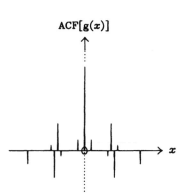

ACF[g(x)]

$$\int\limits_{-\infty}^{\infty} |F(\omega)|^2 e^{i\omega x}\, d\omega = \frac{1}{2\pi} \int\limits_{-\infty}^{\infty} f(y)^* f(x+y)\, dy \qquad (15.8)$$

where the integral on the right gives a formal definition of an ACF. There is always a peak at the origin of the ACF ($x = 0$), because everything correlates with itself (separation = 0). Incidentally, putting $x = 0$ in eqn (15.8) yields a special case of *Parseval's theorem*

$$\int\limits_{-\infty}^{\infty} |F(\omega)|^2\, d\omega = \frac{1}{2\pi} \int\limits_{-\infty}^{\infty} |f(x)|^2\, dx \qquad (15.9)$$

which is sometimes helpful for evaluating integrals if one side is easier to work out than the other.

15.5   Physical examples and insight

Although we began this chapter by thinking about Fourier series as an analogue of the Taylor series for approximating periodic functions, Fourier transforms arise naturally in many branches of science. They occur in both a theoretical context, such as in quantum mechanics, and a physical one, in any experimental situation which entails *diffraction* or *interferometery*. While our discussion has focused on the one-dimensional case, the analysis often need to be generalised to cater for several parameters. This can easily be done by using vector notation, so that eqn (15.4) becomes

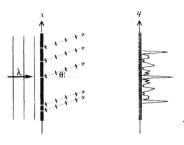

$$f(\underset{\sim}{x}) = \int\limits_{-\infty}^{\infty} \int\limits_{-\infty}^{\infty} \cdots \int\limits_{-\infty}^{\infty} F(\underset{\sim}{\omega}) \exp(i\underset{\sim}{\omega}\bullet\underset{\sim}{x})\, d\underset{\sim}{\omega} \qquad (15.10)$$

and so on, where eqn (15.10) is a surface integral if $\underset{\sim}{x}$ and $\underset{\sim}{\omega}$ are two-dimensional and, a volume one if they are three-dimensional.

To get a better intuitive feel for Fourier transforms, let's consider a few examples from *optics* that should be familiar from school physics. We'll begin by describing the formal set-up of what is technically known as *Fraunhofer* diffraction: a plane wave of light, of wavelength λ, strikes a screen which allows some of it through (usually at well marked slit points) and a pattern of *fringes* is seen to be projected onto a distant wall. If x is the distance along the screen, and q is measured across the wall, then it's not too difficult to appreciate that the net amplitude of all the waves diffracted in a certain direction, $\psi(q)$, is given by

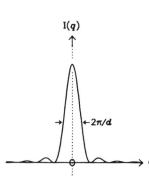

I(q)

←$2\pi/d$

$$\psi(q) = \psi_0 \int\limits_{-\infty}^{\infty} A(x)\, e^{iqx}\, dx \qquad (15.11)$$

where ψ_0 is a constant, $A(x)$ is the *aperture function* which specifies how much light is let through by the screen (it's zero for the opaque regions, and unity at the locations of the clear slits), and q is related to λ and the diffraction angle θ by $q = 2\pi \sin\theta/\lambda$; $\exp(iqx)$ is the part of the plane-wave solution that represents the difference in the path-lengths travelled on going from various points on the screen to a location on the wall. The measured signal, or the number of photons detected, is equal to the intensity $I(q)$

$$I(q) = |\psi(q)|^2 = \psi(q)^* \psi(q) \qquad (15.12)$$

so that the bands of light and dark on the wall are proportional to the modulus-squared of the Fourier transform of the aperture function. Thus, in conjunction with our discussion at the end of section 15.3, the uniform fringes obtained from a *Young's double slit* experiment are mathematically equivalent to $I(q) \propto \cos(qd) + 1$. Similarly, the diffraction pattern given by a single wide slit is related to $I(q) \propto [\sin(qd/2)/(qd)]^2$ or $I(q) \propto [1 - \cos(qd)]/(qd)^2$.

The diffraction patterns of slightly more complicated cases can often be worked out by analysing the situation as a composite of simpler ones which are linked through the convolution theorem of eqn (15.6). Since a pair of wide slits can be regarded as arising from the convolution of a Young's double slit with a single wide one, for example, the resulting bands of light and dark will consist of the product of a set of uniform (cosine) fringes and a sinc function (squared). Similarly, as a short array of evenly-spaced (narrow) slits can be constructed by multiplying an infinitely long 'comb' function and a single wide slit, its diffraction pattern will be made of the convolution of a uniform set of spikes and a sinc function (squared).

Finally, we should note that the measurement of the intensity, $I(q)$ in eqn (15.12), means that information about the phase of the complex Fourier component, $\psi(q)$ in eqn (15.11), is lost. While the importance of the latter can be difficult to appreciate, it is essential if we are to draw any conclusions reliably about the aperture function: $I(q)$ only tells us about the ACF of $A(x)$ in a straightforward manner, as indicated by eqn (15.8), rather than $A(x)$ itself. This translates into such problems as crystallographers have in trying to infer the structure of a protein, say, given the intensities of the *Bragg* spots.

Exercises

15.1 By first showing that sines and cosines are orthogonal, derive the formulae of eqn (15.2) from eqn (15.1).

15.2 For the Fourier expansion of eqn (15.1), derive Parseval's identity

$$\frac{1}{\pi} \int_{-\pi}^{\pi} [f(x)]^2 \, dx = \frac{a_0^2}{2} + \sum_{n=1}^{\infty} (a_n^2 + b_n^2)$$

15.3 A triangular wave is represented by $f(x) = x$ for $0 < x < \pi$, $f(x) = -x$ for $-\pi < x < 0$ and $f(x) = f(x + 2m\pi)$ for any integer m. Show that its Fourier series representation is given by

$$f(x) = \frac{\pi}{2} - \frac{4}{\pi} \sum_{n=0}^{\infty} \frac{\cos[(2n+1)x]}{(2n+1)^2}$$

15.4 Derive the formulae for the intensities of the diffraction patterns from a Young's double slit, with separation d, and single wide slit, of width w.

15.5 Find separately the Fourier transforms of narrow spikes of unit area, or δ-functions, located at: (i) the origin, $x = 0$; and (ii) $x = d$. How do they differ, and what are implications of (only) measuring the intensities?

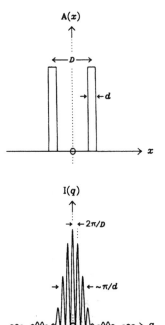

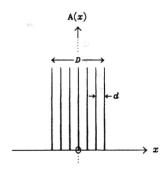

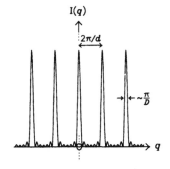

Appendix

Answers to exercises

1.1 (i) 10 (ii) $2\frac{5}{6}$ (iii) 2 **1.2** (i) $ac + ad$ (ii) $a^2 + 2ab + b^2$ (iii) $a^2 - b^2$ **1.3** (i) 8 (ii) $\frac{1}{9}$ (iii) $\sqrt{3}$ (iv) 3 (v) 9 **1.6** (i) 2,3 (ii) $\frac{1}{3}, -2$ (iii) $2 \pm \sqrt{2}$ **1.7** $|k| \geq 4$ **1.8** (i) $(2, -1)$ (ii) $(-1, -1)$ or $\left(\frac{7}{5}, \frac{1}{5}\right)$ (iii) $\left(3, \frac{1}{2}, -2\right)$ **1.9** $32 + 80x + 80x^2 + 40x^3 + 10x^4 + x^5$; $1 + 9x + 36x^2 + 84x^3 + 126x^4 + 126x^5 + 84x^6 + 36x^7 + 9x^8 + x^9$; $x^6 + 12x^4 + 60x^2 + 160 + 240/x^2 + 192/x^4 + 64/x^6$ **1.11** 7/22 **1.12** (i) $1/(x-3) - 1/(x-2)$ (ii) $3/(x-1)^2 + 9/(x-1) - 17/(2x-3)$; (iii) $(3x+5)/(x^2 - 3x - 2) - 3/(x-1)$ **1.13** (i) 1 (ii) $e^{-\beta/2}/(1 - e^{-\beta})$.

2.2 $2y = 5 - x$; $(1, 2)$ **2.4** $\left(-\frac{1}{2}, \frac{3}{4}\right)$ **2.5** $y = x^2 - 4x + 3$; 1,3 **2.6** $y = 0$ when $x = 3, 1$, or -1; when $x = 0, y = 3$; $|x| < 1$ or $x > 3$ **2.9** $(1, -2)$, $r = 3$ **2.10** $\frac{1}{2}$, $\left(\pm\frac{1}{2}, 0\right)$, $x = \pm 2$ **2.11** $y = \pm x$ **2.12** $(x^2 + y^2 + 2x)(x^2 + y^2 - 2x) = 0$.

3.1 $30°, 45°, 60°, \pm 90°, 144°, \pm 120°, 180°, 270°$ **3.2** sin: $\pm\sqrt{3}/2, \pm 1/\sqrt{2}, \pm 1/2, -1/\sqrt{2}, -\sqrt{3}/2, -1/2$; cos: $-1/2, -1/\sqrt{2}, -\sqrt{3}/2, -1/\sqrt{2}, -1/2, \sqrt{3}/2$; tan: $\mp\sqrt{3}, \mp 1, \mp 1/\sqrt{3}, 1, \sqrt{3}, -1/\sqrt{3}$ **3.4** $2t/(1 + t^2)$; $(1 - t^2)/(1 + t^2)$; $2t/(1 - t^2)$ **3.5** $-\pi/3, 2\pi/3$; $-5\pi/6, -\pi/6, \pi/2$; $\pm\pi/2, \pm\pi/3, \pm 2\pi/3$ **3.6** $\pi/12$ or $5\pi/12$ **3.7** $4\sin\theta\cos^3\theta - 4\sin^3\theta\cos\theta$ **3.8** $8\cos^4\theta = \cos 4\theta + 4\cos 2\theta + 3$ **3.9** $\pi/2, \pi/9, 5\pi/9, 7\pi/9$ **3.11** 2.294 Å.

4.1 (i) $-\sin x$ (ii) nx^{n-1} (iii) $-1/x^2$ (iv) $-2/x^3$ **4.2** (i) 0 (ii) $\pi/2 \pm m\pi$, $m = 0, 1, 2, 3, \ldots$ (iii) 0 **4.4** $1/(1 - x)^2$ **4.5** $-1/\sqrt{a^2 - x^2}$ for $0 < y \leq \pi$; $a/(a^2 + x^2)$ **4.6** (i) $6(2x + 1)^2$ (ii) $3/(2\sqrt{3x - 1})$ (iii) $-5\sin 5x$ (iv) $6x\cos(3x^2 + 7)$ (v) $8\tan^3(2x + 3)\sec^2(2x + 3)$ (vi) $(1 - 6x^2)e^{-3x^2}$ (vii) $\ln(x^2 + 1) + 2x^2/(x^2 + 1)$ (viii) $(x\cos x - \sin x)/x^2$ **4.7** $a^x \ln a$ **4.8** (i) $2t/(3t^2 + 2)$ (ii) $[2x - y^2\cos(xy)]/[xy\cos(xy) + \sin(xy)]$ **4.9** (i) min at $x = (1 + 2\sqrt{7})/3$, max at $x = (1 - 2\sqrt{7})/3$, inflex at $x = 0$ (ii) min at $x = -1$ and max at $x = 1$ (iii) min at $r = \sigma 2^{1/6}$ (iv) $r = 2a_0$ [min], $(3 \pm \sqrt{5})a_0$ [max].

5.1 (i) $x^2/2 + \frac{2}{3}x^{3/2} - \ln x + C$ (ii) $\frac{2}{5}x^{5/2} - 2\sqrt{x} + C$ (iii) $2^x/\ln 2 + C$ (iv) $e^{2x}/2 + C$ (v) $\ln(2x - 1)/2 + C$ (vi) $-\cos 2x/2 + \sin 3x/3 + C$ (vii) $\ln(\sec x) + C$ (viii) $x/2 - \sin 2x/4 + C$ (ix) $\frac{1}{2}\ln(1 + x^2) + C$ (x) $\tan^{-1}x + C$ **5.2** (i) $\frac{1}{5}$ (ii) $\frac{3\pi}{16}$ (iii) $9\frac{1}{3}$ **5.3** (i) $\ln[(x - 3)/(x - 2)] + C$ (ii) $-3/(x - 1) + 9\ln(x - 1) - \frac{17}{2}\ln(2x - 3) + C$ (iii) $\frac{3}{2}\ln(x^2 - 3x - 2) + \frac{19}{2\sqrt{17}}\ln[(2x - 3 - \sqrt{17})/(2x - 3 + \sqrt{17})] - 3\ln(x + 1) + C$ **5.5** 8/15; $35\pi/256$.

6.1 $\frac{1}{2} + \frac{\sqrt{3}x}{2} - \frac{x^2}{4} - \frac{\sqrt{3}x^3}{12} + \frac{x^4}{48} + \cdots$ **6.2** 2.8284; 4.1231 **6.3** $-x - \frac{x^2}{2} - \frac{x^3}{3} - \frac{x^4}{4} - \cdots$; $2\left(x + \frac{x^3}{3} + \frac{x^5}{5} + \frac{x^7}{7} + \cdots\right)$ **6.4** (i) a (ii) 0 (iii) $-1/12$ **6.5** 0.406.

7.1 (i) 2 (ii) 3 (iii) $2-3i$ (iv) $6i$ (v) 4 (vi) $-5+12i$ (vii) 13 **7.2** (i) $3+2i$ (ii) $1+4i$ (iii) $5+i$ (iv) $(-1+5i)/2$ (v) $(-1-5i)/13$ **7.3** (i) $\sqrt{13}$ (ii) $\sqrt{2}$ (iii) $\sqrt{26}$ (iv) $\sqrt{13/2}$ (v) $\sqrt{2/13}$ **7.4** (i) $(1,0)$ (ii) $(\sqrt{2},\pi/4)$ (iii) $(1,\pi/2)$ (iv) $(\sqrt{2},3\pi/4)$ (v) $(1,\pi)$ (vi) $(\sqrt{2},-\pi/4)$ (vii) $(1,-\pi/2)$ (viii) $(\sqrt{2},-3\pi/4)$ **7.6** $(1\pm i\sqrt{3})/2$ **7.7** For $n=0,\pm1,\pm2$: (i) $e^{i2n\pi/5}$ (ii) $2^{1/10}e^{i\pi(1+8n)/20}$ (iii) $e^{i2n\pi/5}-1$ (iv) $(-1+e^{-i2n\pi/5})/[2-2\cos(2n\pi/5)]$ $(n\neq0)$ **7.9** $\sin4\theta=4\sin\theta\cos^3\theta-4\sin^3\theta\cos\theta$; $\cos4\theta=8\cos^4\theta-8\cos^2\theta+1$ **7.11** $\cosh(x+y)=\cosh x\cosh y+\sinh x\sinh y$ **7.13** (i) $\cos(\sin\theta)e^{\cos\theta}$; (ii) $e^{ax}[a\sin(bx)-b\cos(bx)]/(a^2+b^2)$.

8.1 (b), (c), (e), (g) **8.2** (i) $(9,5,10)$ (ii) $(-3,-1,-2)$ (iii) $(-13/2,0,1/2)$ (iv) $(0,0,0)$ (v) $(3/2,1/2,1)$ and $(3,2,4)$ **8.3** (i) $(1,2,3)+\lambda(0,1,2)$ (ii) $(3/2,1,2)+\lambda(3/2,1/2,1)$ (iii) $(x-1)/2=z-3, y=2$, and $(x-1)=(z-1)=5(y-1)/2$ **8.4** 5, 6, 24. 43.6°, 21.6°. $(12,0,6)$, $(5,5,5)$ **8.6** $(2,5,-4)$, $(-1,2,-1)$, $(4,10,-8)$. 43.6°, 21.6°. $\underline{r}\times(2,0,1)=(2,5,-4)$ and $\underline{r}\times(5,2,5)=(3,0,-3)$ **8.8** 3, 6, 0 ; $2x+5y-4z=0$ or $\underline{r}\cdot(2,5,-4)=0$; 0 **8.9** 0. No. Yes: $2(1,2,4)-3(2,0,-3)$ **8.11** $1/s$.

9.1 $\begin{pmatrix}5&4\\1&6\end{pmatrix}$; $\begin{pmatrix}-1&-2\\1&-2\end{pmatrix}$; $\begin{pmatrix}6&10\\3&11\end{pmatrix}$; $\begin{pmatrix}9&9\\4&8\end{pmatrix}$ **9.3** $\begin{pmatrix}1&0&1\\1&1&3\\0&1&1\end{pmatrix}$ **9.4** $\lambda_1=0$, $\underline{e}_1=(1,0,-1)/\sqrt{2}$; $\lambda_2=-1$, $\underline{e}_2=(0,1,0)$; $\lambda_3=2$, $\underline{e}_3=(1,0,1)/\sqrt{2}$.

10.1 $3x^2/(1-y)$; $x^3/(1-y)^2$; $6x/(1-y)$; $2x^3/(1-y)^3$; $3x^2/(1-y)^2$ **10.2** $[x^2y^2z^2-1]\sin(xyz)-3xyz\cos(xyz)$ **10.4** $(0,-\frac{1}{4})$; $(-\frac{1}{4},\frac{1}{4})$; $(\frac{1}{4},0)$; $(-\frac{1}{3},\frac{1}{3})$, $-\frac{1}{27}$ **10.7** Saddle: $(0,0)$; minimum: $(\pm a,0)$ **10.8** $\pm(1/\sqrt{2},-1/\sqrt{2})$, $e^{1/2}$; $\pm(1/\sqrt{2},1/\sqrt{2})$, $e^{-1/2}$.

11.1 1 **11.2** $\frac{1+25\sqrt{5}}{120}$; $\frac{1}{2}$. **12.2** $\frac{4}{3}\pi r^3$ **12.4** $\pi/12$

13.1 $\ln(2)/\lambda$ **13.2** (i) $Ax^2=(1+y)/(1-y)$ (ii) $y=-x/(2\ln x+A)$ (iii) $\tan^{-1}[(y+3/2)/(x+7/2)]=A+\ln\left((x+7/2)\sqrt{1+(y+3/2)^2/(x+7/2)^2}\right)$ (iv) $y=(x+A)/\sin x$ (v) $y=Ae^{-x^2}+1/2$ (vi) $y=\sin x+(\cos x+A)/x$ **13.3** $(1-y)/y=Ae^{x^2/2}$ **13.5** (i) $Ae^{-x}+Be^{3x}+(-2\sin x+\cos x)/10$ (ii) $Ae^{-2x}+Be^{4x}+(-8x^2+4x-3)/64$ (iii) $A\cos(\omega_0 x)+B(\sin\omega_0 x)+\cos(\omega x)/(\omega_0^2-\omega^2)$ (iv) $e^{-x/2}[A\sin(\sqrt{3}x/2)+B\cos(\sqrt{3}x/2)]+(1-\omega^2)\cos(\omega x)/(\omega^4-\omega^2+1)+\omega\sin(\omega x)/(\omega^4-\omega^2+1)$ (v) $A\cos(2x)+B\sin(2x)+x\sin(2x)/4$; $[-e^{-x}-2\sin x+\cos x]/10$ **13.6** (i) Ax^{-1} (ii) $(A\ln x+B)x^{-1}$.

14.1 $\sin(xy)+y^2+C$ **14.3** $\psi_{kl}(x,y)=A_{kl}\sin(k\pi x/a)\sin(l\pi y/b)$, where $A_{kl}=\text{constant}$, $k=1, 2, 3, \ldots$, and $l=1, 2, 3, \ldots$. $E_{kl}=\frac{h^2}{8m}[(k/a)^2+(l/b)^2]$ **14.4** $p=\pm1, \pm2, \pm3, \ldots$; $\frac{\tau r}{a}\cos\theta$; $\frac{3\tau r}{4a}\cos\theta+\frac{\tau}{4}(\frac{r}{a})^3\cos3\theta$.

15.4 $\sim[1+\cos(qd)]$; $\sim[1-\cos(qw)]/q^2$ **15.5** (i) ψ_0 (ii) $\psi_0\,e^{iqd}$.

Bibliography

Further reading

The following Oxford Chemistry Primers, Oxford University Press, illustrate the use of the mathematics presented here in a chemistry context.

Atomic spectra. T. P. Softley (1994), OCP 19.

Computational chemistry. G. H. Grant and W. G. Richards (1995), OCP 29.

Quantum mechanic I: foundations. N. J. B. Green (1997), OCP 48.

Thermodynamics of chemical processes. G. J. Price (1998), OCP 56.

Quantum mechanic II: the tool kit. N. J. B. Green (1998), OCP 65.

A book using the mathematics in this Primer, and written in a similar style, explaining probability and statistics in a coherent manner.

Data analysis: a Bayesian tutorial. D. S. Sivia (1996),
    Oxford University Press.

A selection of more advanced books on scientific mathematics; this Primer will serve as a good introduction to them.

Mathematical methods for physicists. G. B. Arfken and H. J. Weber (1995),
    Academic Press, Harcourt Brace & Co.

Advanced engineering mathematics. E. Kreyszig (1998), John Wiley Inc,
    John Wiley and Sons Ltd.

Mathematical methods for the physics and engineering. K. F. Riley,
    M. P. Hobson and S. J. Bence (1998), Cambridge University Press.

Index